ClimateGames

Climate Games

Experiments on How People Prevent Disaster

Talbot M. Andrews

Andrew W. Delton

and Reuben Kline

University of Michigan Press
Ann Arbor

Published in the United States of America by the
University of Michigan Press
Printed and bound by CPI Group (UK) Ltd, Croydon, CR0 4YY
First published March 2024

A CIP catalog record for this book is available from the British Library.

Library of Congress Cataloging-in-Publication Data

Names: Andrews, Talbot M., author. | Delton, Andrew W., 1994- author. | Kline, Reuben., author. | Michigan Publishing (University of Michigan), publisher.
Title: Climate games : experiments on how people prevent disaster / Talbot M. Andrews, Andrew W. Delton and Reuben Kline.
Description: Ann Arbor : University of Michigan Press, 2024. | Includes bibliographical references (pages 179–200) and index.
Identifiers: LCCN 2023050818 (print) | LCCN 2023050819 (ebook) | ISBN 9780472076635 (hardcover) | ISBN 9780472056637 (paperback) | ISBN 9780472904297 (ebook other)
Subjects: LCSH: Climatic changes—Social aspects. | Climatic changes—Effect of human beings on. | Climatic changes—Economic aspects. | Climate change mitigation—International cooperation.
Classification: LCC QC903 .A148 2024 (print) | LCC QC903 (ebook) DDC 363.7/07—dc23/eng/20231127

LC record available at https://lccn.loc.gov/2023050818
LC ebook record available at https://lccn.loc.gov/2023050819

DOI: https://doi.org/10.3998/mpub.12089759

The University of Michigan Press's open-access publishing program is made possible thanks to additional funding from the University of Michigan Office of the Provost and the generous support of contributing libraries.

Contents

Digital materials related to this title can be found on the Fulcrum platform via the following citable URL: https://doi.org/10.3998/mpub.12089759

Acknowledgments

We first and foremost want to thank Matt Lebo and Jeff Segal for putting our team together. This book wouldn't have happened without them. We thank the Stony Brook University College of Arts and Sciences for funding much of the underlying research. We also want to thank all our fantastic coauthors on the experiments throughout this book, including Alessandro Del Ponte, John Barry Ryan, Nick Seltzer, Autumn Bynum, Evgeniya Lukinova, Oleg Smirnov, and Adam Seth Levine. The ideas throughout are shaped by our collaborations. We are also grateful for years of comments and guidance from our friends and colleagues, especially Peter DeScioli, Yanna Krupnikov, Tess Robertson, Markus Prior, and all the members of Stony Brook University's Center for Behavioral Political Economy. Patrick Kraft deserves special mention for providing invaluable technical assistance at a critical moment. John Patty, Katherine Sawyer, and Sean Gailmard also deserve special mention for helping us sharpen our game theory.

We are grateful for the feedback we received from countless people at conferences and workshops. This work benefitted especially from feedback at the Yale School of the Environment, the Harvard Experimental Political Science Graduate Student Conference, the Oklahoma Center for Evolutionary Analysis, and the Arkansas Experimental Economics Seminar (as well as multiple MPSAs, APSAs, and SPSAs). The SESYNC New Scenarios and Models for Climate Engineering Workshop greatly improved our knowledge of climate forecasting and geoengineering. The Penn State Behavioral Models of Politics conference was especially crucial for inspiring this work—it was on the drive back from this workshop that we wrote the first outline.

None of this work would have been possible without the research assistants who helped with the data collection. We thank Daniella Alva, Donovan Bush, Elias Shammas, Hana Kim, Narmin Butt, Payel Sen, Pei-Hsun Hsieh, and Samuel Jens for their hours in the lab.

Talbot wants to especially thank her friends, family, and colleagues who have kindly listened to her talk about climate change and cooperation nonstop, especially Caitlin Davies, Hannah Thorson, Hillary Style, Jacob Martin, Lizzie Atkins, Michelle Io-Low, Scott Bokemper, the Two Crew (Alex Neufeld, Emily Dyer, JT Gwozdz, and Michael Yontz), and Chris, Julie, and Walker Andrews.

Andy cannot begin to express his gratitude to Tess for twenty (and counting) years of advice and love and to his parents Sharon and Jim Delton for everything.

Reuben is eternally grateful to Alegría for the many hours she endured of polemical diatribes about climate change and behavior (though those will surely persist), and to his parents, Lalah and Roger Kline, for supporting him in all that he's ever done.

ONE

Understanding the Challenges of Climate Change

On April 5, 1815, Mount Tambora erupted. Located on the island of Sumbawa (in present-day Indonesia), it would become one of the world's largest volcanic eruptions. The eruption spewed ash and sulfur dioxide miles into the stratosphere. The particles then spread over the globe, cooling the Earth. The Northern Hemisphere experienced a volcanic winter in the summer of 1816, with global temperatures dropping by approximately half a degree Celsius. Twelve inches of snow fell in Quebec City in June. Global agricultural production was decimated. The effects of the eruption were profound, but short-lived. As the sulfur dioxide abated, temperatures slowly rose, returning to normal by 1819.

Fast-forward 200 years. The world's population has been releasing greenhouse gases, such as carbon dioxide and methane, which trap the sun's heat as it hits the planet. Average global temperatures have been rising for decades. Despite decades of warnings and calls to action, little progress has been made in reducing greenhouse gas emissions. Since so little has been achieved through collective action, some have begun to believe that more drastic steps are necessary to keep average global temperatures from reaching a catastrophic tipping point. One approach is a geoengineering technology inspired by volcanic eruptions like Mount Tambora's. The technique is meant to increase Earth's albedo, the amount of solar radiation that the planet reflects. By reflecting more heat back into space, such a plan could partially reverse the steady rise in global temperatures that defined the industrial age. If successful, the effect would be like Mount

Tambora's eruption: large, though temporary. Author Kim Stanley Robinson's best-selling novel, *Ministry for the Future* (2020), opens with a heat wave in India, leading to the death of twenty million people. This event pushes the Indian government to unilaterally deploy technology to manage solar radiation in the face of continued inaction by the international community. But surely no one in the real world would take such a drastic step on their own . . . right?

In 2022, the company Make Sunsets conducted an unauthorized and widely criticized attempt at solar radiation management, using a technique called stratospheric aerosol injection. Aerosol injection involves dispersing sulfuric compounds—typically precursors to sulfur dioxide—high into the atmosphere. Unfortunately, the technology is largely unproven. In fact, there is a significant chance that it could backfire and make some aspects of climate change worse. Scientists have pointed to potential changes in the tropical monsoon cycle, increased acid rain, less efficient solar power, more sunburns, disruption of satellite sensors, and international conflict (Visioni et al. 2018; Simpson et al. 2019; Dagon and Schrag 2016; IPCC 2021).

Make Sunsets' attempt was small enough that it probably didn't have any appreciable effect on the climate. But the company is already selling "cooling credits" to fund future, larger attempts. Even if this gambit, or one like it, does manage to lower temperatures with few ill consequences, no one can be sure how long the effect will last. After all, the massive Tambora eruption only lowered temperatures for a few years. Either way, such measures are only part of the solution: many experts agree that even if methods like aerosol injection are used (and the more we delay other mitigation strategies, the more likely it is that we will have to use such methods), the world still needs to lower emissions and develop measures to adapt to a changing climate (IPCC 2014). At the same time, many experts worry that even talking about this technology will distract both everyday people and policymakers, making them think that cheap solutions can solve everything (e.g., Lin 2013, 2; Hale 2012; Biermann et al. 2022). So, what should the world do?

• • •

This book is about the strategic problems that humanity faces when dealing with climate change. Climate change is a global *social dilemma* (Keohane and Victor 2016). When something is a social dilemma, this means that the benefits of mitigation efforts are collective—the whole world would be better off if all nations reduced their greenhouse gas emissions. At the same time, however, any given nation would be even better off by continuing to emit and letting other nations take care of the problem. Thus, everyone faces a dilemma, a choice between individual interests and collective interests: All nations are best off if all work together, but each one is individually best off from business as usual. In this way, the prevention of dangerous climate change is a problem involving all of the world's billions of inhabitants. Whether a lone individual, nonprofit, or country should use solar radiation management is just one piece of the puzzle.

Mitigating climate change is complicated by other strategic problems: Many aspects of the climate and climate change are not fully understood, creating uncertainty over what to do about the problem. Climate decisions are made by leaders who could have different knowledge or goals than everyday people. Those wielding the most power to fight climate change are developed countries, but many of the countries most at risk are in the developing world. And many pieces of the puzzle are intergenerational; what we do will largely affect future generations, not ourselves. The focus of this book is to probe these strategic problems, to see how real people understand them and react to them.

To do so, we turn to *economic games*. Economic games are stripped-down representations of strategic problems that we either face now or could face in the future. As we'll see throughout this book, our games involve groups of people working together, with real money on the line, to prevent simulated disasters from wiping out their earnings. Economic games are in some ways like table-top strategy games, such as Risk or Monopoly. Players are given clear rules that allow them to compete and cooperate with each other. Like board games, economic games are limited only by the designers' imaginations. Economic games allow researchers to create little hypothetical worlds in the lab. Although they involve money, they are not only about money. Economic games allow researchers to see what people value: money, yes, but also things like fairness, status, reputation, reciprocity, and the welfare of others.

As Chapter 2 describes more fully, economic games have many features that make them ideal research tools for studying problems of strategy. They have clear material incentives, they are simple compared to the real world, and they are fully transparent. These features allow researchers to know exactly what is happening in the game. Though games sacrifice some realism, they make it much easier to learn how people solve strategic problems than could be done by studying the real world directly. Politics in the real world is messy. People's actions are influenced by a bewildering jumble of institutions, laws, social networks, incentives, beliefs, and much else. So, learning exactly why a person, a company, or a nation took a particular action is hard to do with much confidence. With games, we the researchers know all of the material incentives people face—because we designed them. Accordingly, we also know which incentives people do not face, and which features of the environment are not present. This aspect of experimental control allows us to rule out alternative potential causes and to focus on a small number of factors. The starkness of the experimental environment is a feature, not a bug, and a powerful feature at that.

We and many other researchers have used what we call the *disaster game* to study how people might prevent disasters when working in groups (Milinski et al. 2008). In this game, players working in a group (perhaps strangers over the internet or students in a lab) are each given money. When the game begins, the players stand a good chance of losing their money due to simulated disaster. But if the group as a whole contributes enough, they can mitigate disaster, preventing it from wiping out whatever money they have left. It sounds simple, but as we'll see, this basic game can be adapted with endless variations. This allows us to simulate some of the strategic problems noted above and see how people react in the controlled and transparent miniworlds of games.

To see the reach and range of even this one game, let's briefly preview just a sample of the variations we and others have created. In Chapter 3 we use the disaster game to see how people respond to different technologies to stop climate change. When would you be willing to invest in high-risk but potentially high-reward technology, like many forms of geoengineering? When would you play it safe and invest in something tried and true? In the games here, players have to decide not just *whether* to contribute but *how* to contribute.

In Chapter 4 we use the disaster game to see whether people will help *others* avoid disaster. Here, when groups contribute, they can only help an altogether different group. Would you be willing to pay to help others?

In Chapter 5 we use the disaster game to test a related question: Will people avoid *creating* disaster for others? Here, the cost of mitigation is controlled by the players themselves. If they can restrain themselves from producing too many simulated emissions, the costs of mitigation are low; if not, the costs spiral out of control. You might restrain yourself if you have to mop up your own mess, but is self-control possible when others have to clean up after you?

In Chapter 6 we use the disaster game to study tensions between *leaders and citizens*. When would you trust a leader to give you good information about the costs of mitigation? Would a leader trust you to react rationally to their policies?

This list just scratches the surface of what the disaster game has been used for. Other studies involve inequality, economic development, voting, solutions that could explode in your face, uncertainty about how awful disaster will be, clear thinking and muddled thinking, and much more. We'll also meet real island nations as they struggle to deal with rising seas, presidents skeptical that climate change is happening at all, abalone fishers trying to keep their stocks sustainable, and scientists who think resurrecting mammoths can help solve climate change.

From this point on, readers can use this book in whatever way best suits their interests and knowledge. If you want more detail on the real-world problems we'll investigate, stick around in this chapter. We'll introduce you to a variety of real dilemmas and complications involving climate change. As we'll see, climate change is not one single problem, but a tangle of related problems. If you don't know much about games, in the next chapter we provide a nontechnical introduction—why researchers use games, how they create them, and what their strengths and weaknesses are. And if you just want to explore particular strategic problems and already understand games, you can skip to any other chapter that calls to you. Each one is written to be largely self-contained. (Though if you have never read about the disaster game before, we recommend reading the beginning of Chapter 3.) Although games often have complicated math models standing behind them, we want this book to be accessible to researchers of all kinds, to students, to policymakers, and to anyone else who is interested. Thus, we focus on verbal intuitions for the games and have kept math to an absolute minimum, using just basic arithmetic to illustrate a few key points. Readers who want the gory details can consult the original papers and technical appendices.

The Strategic Challenges of Climate Change

Climate change may be a social dilemma, but so are many other problems: creating a clean water supply, maintaining law and order, building roads and infrastructure, providing national defense, and others besides. All of these make communities or nations better off as a whole, but any individual would be even better off enjoying the benefits without doing their part to contribute. What makes climate change special is that it involves a constellation of additional challenges layered on top of the basic problem of a social dilemma. There are four we address in this book: uncertainty, making decisions that affect others, dealing with self-created disasters, and tensions between elites and laypeople.

Challenge 1: Uncertainty All the Way Down

Climate change is real and is caused primarily by humans emitting greenhouse gases such as carbon dioxide (CO_2) into the atmosphere. But climate change is also surrounded by uncertainty. How bad will things get if we fail to stop climate change? How should we define climate change tipping points, the point after which particular climate disasters are certain? How much do we need to reduce carbon dioxide emissions to stop climate change?

Let us be clear on exactly what we do and do not mean by climate change uncertainty. We do not mean uncertainty manufactured by political elites. Climate change uncertainty is not Oklahoma senator Jim Inhofe holding a snowball on the Senate floor, asking how climate change can be real when it's snowing (Bump 2015). It is not Utah senator Mike Lee saying the Green New Deal is just as fanciful as a picture of Ronald Reagan riding a velociraptor, a picture he showed on the Senate floor for emphasis (Chiu 2019). It is not President Donald Trump tweeting "Brutal and Extended Cold Blast could shatter ALL RECORDS—Whatever happened to Global Warming?" (Trump 2018).

Instead, we mean real scientific uncertainty around how to best mitigate climate change and the impacts of failing to do so. Indeed, assessing and communicating uncertainty is a central concern of the Intergovernmental Panel on Climate Change (IPCC), one of the most important global authorities on climate change. It regularly releases reports that synthesize the state of the art on climate change. The IPCC explicitly characterizes uncertainty by describing the evidence along three dimensions: the amount of evidence, the quality of the evidence, and to what degree the evidence points in the same direction (IPCC 2010).

For example, the IPCC is highly confident that tropical cyclones will produce more precipitation if global temperature rises 2°C compared to 1.5°C. A large body of research has tested this claim, and after years of investigation the studies have consistently found that higher temperatures go along with more intense cyclones. At the same time, there is low confidence that precipitation outside of tropical cyclones will increase if global temperature rises 2°C. Our question for Chapter 3 is: What does this type of uncertainty mean for behavior?

Uncertainty in Climate Change Impacts

According to the newest IPCC assessment report, things are certainly going to get worse if we don't reduce carbon emissions. But in what ways will things get worse? And by how much? Some impacts are virtually certain, such as the seas rising. Other impacts are only reasonably likely. For example, we will probably experience more extremes in precipitation. This means sometimes getting very heavy storms and sometimes severe droughts. As global temperature rises, we are also likely to experience more heatwaves. The IPCC additionally has some confidence that wildfires will become worse as temperatures increase (IPCC 2021).

Although there is some uncertainty around *whether* we will experience more heatwaves and changes in precipitation, more serious uncertainty surrounds *where* and *when* we will experience these events (IPCC 2021). The severity of these events depends not only on the physical characteristics of the event, but on what countries are affected and how prepared people are as the events take place.

Take, for example, the case of really, really hot weather. On June 19, 2021, both Portland, Oregon, and Phoenix, Arizona, experienced temperatures over 110°F. In Portland and nearby neighborhoods, fewer than half of the people had air conditioning, and thousands lost power because of the heat (Baker 2021). Public streetcars had to close as the tracks and power lines buckled under the extreme temperatures (Live 2021). As a result of the heat and lack of infrastructure to prepare for the heat, over 600 people died that week (Baker and Olmos 2021). In Phoenix, Arizona, temperatures exceeding 110°F didn't even make the local news. (One of us grew up in Phoenix and can personally attest that such temperatures are common and forgettable; another of us grew up in Portland and can attest that they absolutely are not!)

Of course, this is an extreme example of how different regions may be more or less prepared to handle extreme weather. But it illustrates that uncertainty over where and when is bound up with uncertainty over how people will *experience* an extreme weather event. And indeed, while we will see more extremes in heat and precipitation, the location of these extremes will depend on features of the local atmosphere (IPCC 2021). Besides straightforward effects like droughts, climate change can potentially push people into violent conflict. It is not only possible that people will fight over scarce resources, more indirectly scarcity can decrease the capacity of a state to prevent violent conflict (Homer-Dixon 1999).

Climate change may increase scarcity through droughts, floods, heatwaves, and other disasters. Some work suggests that disasters like these do indeed create outbursts of violence (Nel and Righarts 2008). At the same time, the places most likely to experience these disasters tend to have weaker states that cannot easily handle them. So, do climate disasters actually cause violence? Or is violence breaking out because states where disasters happen are weak, but the disaster itself is irrelevant? It is extremely difficult to tell what is causing what when it comes to disasters and conflict (Gleditsch 2012; Slettebak 2012), and most work has focused only on conflict within a country while ignoring interstate conflict or one-sided violence like genocide (Gleditsch 2012). The IPCC has little to say about climate change and social issues such as violence. As much uncertainty as there is in the relationship between climate change and physical systems, this uncertainty is only amplified when trying to understand how these physical disruptions affect people.

Uncertainty in Climate Change Prevention

We need to reduce carbon emissions to mitigate climate change, and thankfully there are already many tools for this. The IPCC stresses the need to shift toward more energy efficiency, to change our infrastructure and mass transportation, and to increase reliance on renewable energy sources like wind and solar power (IPCC 2018). It is certain that all of these strategies will help reduce the amount of carbon dioxide emitted into the atmosphere. However, it is increasingly unlikely that these certain, incremental technologies will be enough to avoid dangerous climate tipping points.

The IPCC describes many pathways through which we successfully avoid the most severe impacts from climate change. Importantly, any successful pathway relies on geoengineering, specifically the subcategory of geoengineering called carbon dioxide removal. Carbon dioxide removal does *not* reduce the amount of carbon dioxide we emit, unlike transitioning to emission-free energy. But it can remove carbon dioxide from the air or otherwise prevent the downstream effects of emissions. These strategies include methods as simple as afforestation (i.e., planting more trees) or more complex technologies like "bioenergy with carbon capture and storage" in which we produce electricity by burning biomass but capture the resulting pollutants. These strategies all demand difficult tradeoffs; for example, afforestation competes with agriculture for land use. Further, they introduce two types of uncertainty.

First, many of these technologies are uncertain in that they simply don't yet exist, at least on a scale where they could successfully stop climate change. Money and time invested in these new projects could be time wasted. The IPCC has considered multiple "negative emissions" technologies, including directly capturing carbon from the air, alkalizing the oceans, or sequestering carbon in the soil. Some of these technologies will probably need to be part of the solution, but it's unclear which ones.

Second, many of these strategies could have negative side effects. Take again the case of stratospheric aerosol injection. Though not a negative emissions technology, aerosol injection is a type of geoengineering that could reduce the rise in earth's temperature (IPCC 2021). Nonetheless, aerosol injection is uncertain because it is simply not ready for wide-scale deployment. And if it were deployed, it has potential downsides. It could disrupt precipitation patterns around the world, including disrupting the Asian and African summer monsoon seasons, making some populations more at risk for starvation (IPCC 2021). It also has the potential to decrease rain in the Amazon, or exacerbate ocean acidification (Visioni et al. 2018; Simpson et al. 2019; Dagon and Schrag 2016; IPCC 2021). Finally, the effects of solar geoengineering are only temporary, and it's possible that as the effects fade, there will be a rebound to *increased* warming (IPCC 2021). Exactly how much aerosol injection would help or hurt is still being debated.

Climate change mitigation is uncertain because we will need to rely on technologies that may not exist yet. How do we know where to invest time, money, and research to best solve the problem? This question is complicated by uncertainties about the effects of these technologies. Even if they reduce the rise in temperature, is this offset by their side effects? In Chapter 3 we focus on these uncertainties, in both how to mitigate climate change and what happens if we fail to do so. Can people avert disaster even when they do not know exactly what to do or exactly what will happen?

Challenge 2: Deciding for Others

On January 14, 2022, the underwater volcano Hunga-Tonga-Hunga-Ha'apai erupted, sending ash and steam up into the atmosphere. But unlike Mt. Tambora, the problem with this eruption wasn't just the volcanic cloud sent into the air. An additional threat was the tsunami. The earthshaking force of the eruption sent waves around the world, including four-foot waves that crashed into the nearby island of Tongatapu, Tonga (Regan 2022). Buildings along the coast were destroyed and several people died.

This disaster was made worse by climate change. While rising global temperatures do not affect volcanic eruptions, they do cause sea levels to rise. This happens, first, because of melting land ice, like the Greenland ice sheet, running into the ocean. This happens, second, because as liquid warms, it expands. You might not notice this if you reheat your coffee—your coffee does expand when warmed, though not enough to tell in a small mug. But even a little expansion, multiplied by the size of the entire ocean, is a major contributor to sea level rise (Lindsey 2020). Rising sea levels are a problem on their own for low-lying island nations like Tonga. However, they are especially problematic when combined with an event like a volcanic eruption. Rising sea levels only make islands more vulnerable to tsunamis, and to other hazards like hurricanes and cyclones.

Tonga is paying huge costs because of climate change—they are only now beginning to recover from this recent disaster—and they are not alone. Other small islands, like Kiribati, are facing similar struggles as they sink into the ocean. Yet these islands' inhabitants are not responsible for the problem. All Pacific Island nations combined account for only 0.03% of global carbon emissions. This is not just because the islands hold only a tiny fraction of earth's population. Pacific Islanders also emit less per person: The average person in Tonga only emits 1.3 tons of carbon dioxide each year, compared to 15.2 tons by the average person in the United States (2020). Even if Tonga cut their carbon emissions to zero, the effect of global emissions would be miniscule, probably ignorable. The crux of the matter is that Tongans and other Pacific Islanders are relying on the rest of the world to make decisions about climate change, while they pay the costs of slow mitigation. Our primary questions in Chapter 4 examine how people make decisions for others in the face of disaster. When Tonga faces disaster, can the rest of the world cooperate to help? And, more broadly, when some people are more at risk but less able to help stop climate change, can we still successfully cooperate?

Voting for Disaster Prevention

Of course, making decisions for others in the face of climate change isn't limited to national leaders arguing about climate policy. Everyday people also decide for others. Let's look at another example, the case of Oregon wildfires. When you think of Oregon, you probably think of lush Pacific Northwest forests and lots of rain. But before the year 2001, on average 198,000 acres burned every year across the state. Since 2001, that number has more than doubled to 433,541 acres per year (Urness 2021). This increase has many causes, ranging from changes in forest management policies to increased building in fire-prone areas. But it is also exacerbated by climate change, which enhances two key ingredients for bad wildfires: heat and droughts.

The plurality of Oregon voters live in Portland, Oregon, a city largely safe from direct damages from wildfires. While they do not directly experience the damages, Portland voters have to make decisions that will affect those who are more vulnerable. For example, in 2018 voters had to decide between Knute Buehler and Kate Brown, both running for governor. Buehler, the Republican candidate, opposed climate policies such as cap-and-trade programs. He did support wildfire mitigation and adaption policies that were more business-oriented. And he advocated for policies to encourage businesses to invest in water infrastructure to help mitigate droughts. He also advocated for the timber business and wanted to empower them to manage forests to prevent wildfires. Brown, on the other hand, was a strong advocate for cap-and-trade climate change mitigation policies. She also had wildfire prevention as a top priority, advocating for funding for firefighters and first responders to help respond to the disaster. And she pushed for policies to thin forests on government land, to slow the spread of fires.

Although Portland voters are not in direct danger from wildfires, they make up a large portion of the gubernatorial electorate, swaying the election one way or another. How then should they vote on wildfire policies? Some theories predict they shouldn't even bother to get informed about the issue, since it doesn't affect them at all (Downs 1957). Others predict they might try to get out and vote to help those affected by the wildfires (Dawes, Loewen, and Fowler 2015; Andreoni 1995). In Chapter 4 we ask whether people will try to help others avoid disaster and whether they will do so effectively. For instance, in a complicated issue like wildfires, will people not only take the time to vote, but take the time to inform themselves about the issue and possible solutions? (The voters ultimately choose to elect Kate Brown. She has since passed Senate Bill 762, which provides $220 million for Oregon's wildfire preparedness; see KGW 2021.)

The example of choosing between governors highlights another element of climate change mitigation. While individuals can push for certain policies or take steps on their own to reduce carbon emissions, many of the big decisions about climate change will be left up to the political elite. Will people elect leaders who push for mitigation and disaster prevention? Some people don't believe in climate change, but holding them aside for a moment there is still variation in climate attitudes. Some support rapid transitions to carbon-free energy, strict carbon taxes, and other regulations to slow the warming of our planet. Others prefer leaders who are hard bargainers, pushing other nations to pay the cost of decarbonization. In Chapter 4 we look closer at what kinds of leaders people prefer, and how even the basic process of getting to choose who represents you in climate negotiations changes mitigation outcomes.

Challenge 3: Self-Created Disaster

Climate change is a unique problem because, in part, it's a problem we've created ourselves. While we don't have control over natural hazards like volcanic eruptions or earthquakes, humans have changed the climate by emitting greenhouse gases into the atmosphere. And we further exacerbate the damages by not adapting effectively.

Take New York City. The Big Apple is threatened by rising sea levels, more flooding, and more severe and frequent storm surges. Even if we cut all carbon emissions today, the climate would not immediately heal from the damage we've done. So, we also need to adapt to the changes and risks we've already created. One solution, proposed by the Army Corps of Engineers, would be for New York City to build a sea wall around the city, letting ships in and out but closing in advance of storm surges. Yet this hasn't happened. Some of the reasons are better than others—it would cost over $100 billion, it wouldn't protect the city from flooding from high tides, and city residents complain it would be ugly (Barnard 2021). But there may also be something about this being a disaster of our own making that makes it especially challenging. This is what we tackle in Chapter 5.

Looking to the Past and Deciding for the Future

Climate change is a self-created problem, but the consequences span generations. Those alive now are paying a cost generated by those who came before. We can get records of both global temperature and carbon dioxide concentrations from things like ice cores, taken from deep inside Arctic glaciers (Stoller-Conrad 2017). These records, combined with other sources, indicate that global temperatures started rising as early as the 1830s when the Industrial Revolution began the transition to industry powered by enormous amounts of fossil fuel. In most places, carbon emissions have only continued to rise. In the United States, total carbon emissions have gone from approximately 650 million tons in 1900 to 4.7 billion tons in 2020—a more than sevenfold increase.

Just as we are dealing with problems created by generations long gone, we are pushing problems onto future generations. The IPCC formalizes different possible futures, ranging from those defined by drastic cuts to carbon emissions and investment in infrastructure, to those characterized by business as usual. Cutting carbon emissions now is expensive but will prevent future catastrophe for our future selves and future generations. In Chapter 5 we ask if people can limit their emissions even when the consequences of their decisions are passed on to those in the future or to people in other countries. Can people restrain themselves?

Who's Responsible?

Of course, while humanity has collectively changed the climate, the responsibility is not evenly shared. The United States is responsible for the most carbon dioxide emitted in total since 1850, having emitted nearly twice as much as the country in second place (Evans 2021). And there are great disparities in how much each country has emitted historically—even accounting for population size. These disparities are tied to industrialization. As countries industrialized, they grew in wealth but also emitted more.

This creates a tension: to alleviate historic inequalities in development and to gain a decent standard of living, some countries must be allowed to further develop, meaning they will emit more and more per capita. To avoid dangerous rises in global temperature, wealthy countries would therefore need to bear the brunt of the costs for mitigation or adaptation. Unfortunately, it's hard to imagine wealthy countries signing up for that deal. Indeed, concerns about shouldering too much of the weight are one reason why the United States refused to sign the Kyoto Protocol, an international agreement to help coordinate climate action. In Chapters 4 and 5 we examine this tension: Under what conditions will rich actors help poorer actors avoid disaster?

Challenge 4: Tensions between Policymakers and the Public

Individual actions alone cannot achieve climate change adaptation and mitigation . Governments need to play a role, for example by subsidizing clean energy, investing in infrastructure for rising sea levels, or providing relief in the wake of extreme weather. In some ways, this is necessary—not everybody can be an expert in everything, climate change or otherwise. Most people have other things on their minds besides creating detailed plans about the climate (Scruggs and Benegal 2012; DeSombre 2018).

Policymakers and political elites therefore must deal with the details. Thankfully, they now have more resources to do so. The newest IPCC report, coming from just one of three working groups, is 3,949 pages long (IPCC 2021). Even the short summary for policymakers is 40 pages, synthesizing dense academic literature on the potential consequences of climate change. This report draws from a multitude of scientific disciplines, bringing in evidence from computational modeling, chemistry, ecology, atmospheric sciences, and others. In short, it is not a document your average person is going to read. So, it is up to policymakers and their advisors to read information like this and implement policies to effectively address the climate problem.

Can We Trust Our Political Leaders?

In an ideal world, policymakers would be able to synthesize complex information about the causes and consequences of climate change, as well as the available solutions, and then implement effective policies for mitigation or adaptation. But of course, we do not live an ideal world. We have the technology available to us to successfully mitigate climate change; for example, we have the capacity to transition away from fossil fuels and toward renewable energy sources. We even have technical solutions to buy time to make this transition—we could rely more on nuclear energy, or invest in more negative-emissions technologies (IPCC 2021).

These policies have not been put into practice, for many reasons. For example, even well-meaning policymakers have to balance considerations beyond mitigation and adaptation. Rapid decarbonization has the potential to damage the economy (IPCC 2014), and the transition away from nonrenewables has uneven impacts. In West Virginia alone, the number of people working in coalfields has declined from 100,000 in the 1950s to fewer than 20,000 by 2020, decimating the local economy (Scheuch 2020). Many renewable energy sources require a lot of land to be successful, putting renewable energy in competition with farmland. As discussed above, negative emissions technologies might be necessary, but they also have potential drawbacks. So, even the most well-meaning policymaker doesn't have a straightforward path to climate change mitigation or adaptation.

Further, not every policymaker is well-meaning. Some flat out deny the existence of human-caused climate change. Mo Brooks, an Alabama congressman, blamed sea level rise on coastal erosion rather than climate change. Don Young, a representative from Alaska, called climate change a "scam." Senator Marco Rubio argued that "there has never been a time when the climate was not changing." Indeed, as of 2019, at least 130 members of Congress had publicly doubted or denied climate change (Cranley 2019).

Now, imagine you're an American voter deciding how to vote in an upcoming election. Also, imagine you care deeply about stopping climate change (which, since you're reading this book, is probably not too much of a stretch). It's easy then not to vote for the policymaker talking about how they don't even believe in climate change!

Unfortunately, policymakers can deviate from our best interests in more insidious ways, even while appearing to care about climate change. Take, for example, spending on flood prevention. Climate change exacerbates both coastal flooding due to rising seas and inland flooding from extreme precipitation. More and more money is therefore being devoted to flood prevention. In 2020, Virginia's government set aside $45 million for flood mitigation, and Texas set aside nearly $800 million (Hersher 2020). While expensive, spending on prevention like this is much more efficient than spending to rebuild after a disaster. Yet, the public generally supports the opposite, preferring relief spending over prevention spending (Healy and Malhotra 2009). This might be in part because they worry that representatives can benefit themselves at the public's expense. When policymakers have funds for the general purpose of flood prevention, but not committed to specific projects, policymakers can abuse the flexibility (Gailmard and Patty 2018). This is what happened in Oklahoma when the public approved a $25 million bond to rebuild dams and prevent flooding. Over $4 million was instead used to move properties connected to city leaders in the town of Kingfisher—the same leaders who proposed the bond (Dillion and Cross 2013). While technically this did protect properties from flooding, it fell far short of the stated goal of the policy. Instead of improving local infrastructure and protecting the public from increased flooding, it safeguarded the assets of the policymakers.

Voters seeking to support effective mitigation and adaptation can easily vote against policymakers who deny the existence of climate change. It is more difficult to identify the most effective mitigation and adaptation policies, and to tell when policies might mostly enrich leaders. In Chapter 6 we ask: Can people figure out which policymakers and policies to trust?

Can Our Elites Trust Us?

While everyday citizens might struggle to identify which elites to trust, concerned policymakers similarly struggle to mobilize support for climate change policies. In 2019, only 72% of adults in the United States believed climate change is happening, and only 60% believed Congress should do more to address the issue (Leiserowitz et al. 2019). Well-meaning policymakers who want to help stop climate change can run into problems when the people who vote for them don't believe in or don't care about climate change.

Thankfully, there are ways to increase belief in climate change and support for mitigation. People look to their elected officials when it comes to climate change. If voters' congressional representative express support for mitigation, those voters are more likely to do so as well (Carmichael and Brulle 2017; Brulle, Carmichael, and Jenkins 2012). When the media share information about climate change, especially related to extreme weather events like hurricanes or floods, people are more likely to believe in climate change (Roxburgh et al. 2019; Brulle, Carmichael, and Jenkins 2012). And when people themselves experience extreme weather made worse by climate change, they also become more concerned about the issue (Howe et al. 2019).

People also respond to leaders and their policies. For instance, voters reelect policymakers who successfully pass relief spending (Healy and Malhotra 2009). However, some voters may respond inefficiently to climate policy. If everyday voters are overly optimistic that one approach alone is sufficient to stop climate change, they might withdraw support for other, still necessary, mitigation strategies (for example, see Amundson and Biardeau 2018; Barrett 2007; Bodansky 1996; Hale 2012; Kolbert 2014; Lin 2013; Reynolds 2015; Scott 2012). We probably need some geoengineering to prevent dangerous changes in the climate. But geoengineering alone is not the whole solution. Yet if voters believe it is, they may not want elites focusing on solutions like renewable energy. The upshot is that if voters believe geoengineering is a silver bullet, and policymakers deploy these technologies, we might actually be worse off. In Chapter 6 we ask a twin set of questions: Are people overly optimistic about apparent silver bullets? Do elites expect too much optimism from the public and avoid apparent silver bullets in response?

• • •

In the next five chapters, we are going to dive deeper into economic games, using them to test whether people can navigate these challenges. Of course, climate change is a daunting problem, and our decisions now shape the dangers we face in the future. But what we find should inspire some confidence. Despite many roadblocks, people are consistently generous and effective problem solvers. They cooperate to prevent disaster.

TWO

Creating Worlds in the Lab

We, your authors, are experimenters who study people. We do not study how air circulates and creates hurricanes or cyclones, we cannot design a machine that removes carbon dioxide from the air and stores it underground, and we cannot evaluate economic proposals about whether to invest in solar cells or wind farms. So what use are we? What we can add to the conversation is to uncover *how people think about climate change*. What solutions do they find compelling? Are people willing to work on behalf of others to prevent disaster? Can elites and citizens with different knowledge and different incentives cooperate successfully?

Scores of social scientists study how people think about the climate. One method is polling the public. For example, Pew Research interviewed a representative sample of Americans to ask them about climate change. In their sample, 79% of people believed humans have a "great deal" or "some" responsibility for global warming. But Democrats and Republicans differed sharply, at least among people whose ideology matched what is typical for their party: 96% of liberal Democrats believed climate change is partially human caused while only 53% of conservative Republicans did so. The difference was still apparent but somewhat reduced for other partisans: 91% among moderate or conservative Democrats and 77% among moderate or liberal Republicans (Funk and Hefferon 2019).

Many pundits argue that people don't believe climate change is happening and aren't concerned about it partly because they don't know enough about it. Daniel Kahan and colleagues tested this by polling Americans (Kahan et al. 2012). They asked people questions that gauged their scientific and numerical literacy, such as: "True or false: Antibiotics kill viruses as well as bacteria." (Answer: false.) "A bat and a ball cost $1.10 in total. The bat costs $1.00 more than the ball. How much does the ball cost?" (Answer: $0.05). Kahan and team counted people as ignorant if they got these questions wrong. Are ignorant people especially likely to reject climate change as a threat? Nope. Ignorant people were no more or less likely than informed people to declare that "climate change poses [a risk] to human health, safety, and prosperity." Instead, worry about the risks of climate change was mostly confined to people with "communitarian" views on social order, rather than "individualistic" views. Consider this statement: "Government should put limits on the choices individuals can make so they do not get in the way of what is good for society." Communitarians tend to agree with this sentiment while individualists disagree. Polling the public helps us learn which beliefs and opinions relate to concerns about climate change.

Other scientists study people's thoughts on climate change with survey experiments. In a typical survey experiment, people read one of several short passages, sometimes written by the experimenters and sometimes taken from a newspaper or magazine. After reading their passage, a research subject answers questions like whether they believe what it said or whether they support possible actions the government could take. In one survey experiment by Toby Bolsen and Jamie Druckman (2018), the researchers randomly assigned some people to read that most scientists believe climate change is real and partly caused by humans. Other people were not given this information. The researchers found that people who read that there is a scientific consensus were more likely to believe that, yes, there is a consensus. However, if people were randomly assigned to also read that climate change research might be politicized, reading that advocates might be "selectively using evidence," then learning that there is a consensus no longer increased belief in a consensus. The benefit of experiments is that researchers learn *what causes what*. This is because random assignment ensures that the *only* thing varying across different sets of people is (in this example) the passages they read. (This is why in medicine randomized trials are the gold standard for testing whether a new drug works.) So, if different groups of people report different beliefs in an experiment on climate change, it must be because of what the researcher manipulated, not preexisting differences between the people.

Polling and survey experiments are useful to understand how people think about climate change, and we have used them ourselves (Andrews et al. 2022; Andrews and Smirnov 2020; Levine and Kline 2017; Simpson et al. 2021). However, we believe the approach we take in this book—economic games—has virtues as well. The rest of this chapter explains what games are and why they are useful.

Economic games are ways of creating strategic interactions between people with real stakes on the line—and that are tractable for scientists to analyze. To simplify a bit, conducting research using games consists of three steps. First, a researcher identifies a problem they want to understand. Second, they come up with a laboratory version that captures the strategic elements they want to study—the researcher constructs a game. Third, the researcher develops a theory to explain how people should or actually do behave. (This is often, but not always, a mathematical theory.) We'll give many examples of these steps throughout this chapter.

An actual boardgame like Risk or Monopoly is a game in the technical sense (so long as there is something at stake for winning, like money, fame, or bragging rights). What makes something a game as we use the term is that the best course of action for *you* depends on what *other people* are doing. If you ignore where an opponent is building up their troops in Risk, you're likely to be conquered. If you allow other players to buy Boardwalk and Park Place in Monopoly, you'll probably go bankrupt. Although fun to play, boardgames are not great research tools. The problem is that they are so complex that researchers cannot suss out why players made the choices they did.

The same difficulty of teasing things apart is also true in real-world games of politics and economics. When political leaders jockey over an arms-control deal, they are playing a game (a very serious one!). Unfortunately for researchers, we do not always know the incentives real leaders face or the information they are privy to. A president knows classified information about the military that informs his decisions. But these secrets are unavailable to a political scientist when she tries to understand the president's choices. Laboratory games surmount all these problems at once. Researchers create the games, which means we have complete knowledge about the material incentives, information, and choices available to the people playing them. And laboratory games are (relatively) simple, so they are tractable to analyze. Let's take a look.

Decisions, Decisions

You might recognize the names of Daniel Kahneman and Amos Tversky. Kahneman won the Nobel Prize for the pair's decades-long collaboration on psychology and economics. Tversky would have shared the honor, but he passed away before it was awarded. They are best known for two related accomplishments. One was to document many ways that human intuition seems to fall short of mathematical and logical rigor (Tversky and Kahneman 1974, 1986). For instance, people might read about Linda: "Linda is 31 years old, single, outspoken, and very bright. She majored in philosophy. As a student, she was deeply concerned with issues of discrimination and social justice, and also participated in antinuclear demonstrations." Which of the following is more likely to be true of Linda? (1) "Linda is a bank teller" or (2) "Linda is a bank teller and is active in the feminist movement." Many people say (2) is more likely. Yet, according to Tversky and Kahneman, logic demands that (1) must be more likely: Feminist bank tellers are necessarily a *subset* of all bank tellers. So, it must be more likely that Linda is just a bank teller than both a bank teller *and* a feminist (but see Hertwig and Gigerenzer 1999 for why answering (2) is best in the context of everyday conversation).

Their second accomplishment was to create a theory that described how people make "decisions." We put decisions in scare quotes because it has a technical meaning here. Decisions involve a *single* person deciding between two or more alternatives. Often, there is some risk involved. Consider this decision: Do you want $100 for sure or a 50% chance of $250? (Perhaps if you pick the latter option, we'll flip a coin to determine your winnings.) If you pick $100, you get the $100. If you pick the gamble, then you would expect, on average, to get $125. If you only cared about the so-called "expected value" of your choice, then you should take the gamble—$125 is bigger than $100. But maybe you do not like to gamble or do not want to walk away with nothing. In that case, you might choose the sure $100. There's no right or wrong answer to this problem; it depends on your goals and values.

Although scientists cannot necessarily tell you which choice is best in a decision like this, they still want to describe how decisions are made. The reigning theory in economics and psychology, prior to Kahneman and Tversky, was *expected utility theory*. At the risk of oversimplifying, expected utility theory holds that people assign a subjective, personal value—called a "utility"—to each potential outcome, rather than take those outcomes at their objective value. They then multiply these utilities by the probability that they will occur; this is the "expected" part of the theory's name. Whichever option has the largest expected utility is the one chosen (for accessible introductions, see Glimcher 2011, Hastie and Dawes 2009). (Proponents of this theory do not necessarily argue that these computations are conscious. In fact, some supporters of expected utility theory do not believe it's a theory of cognition at all. Instead, it's merely an as-if theory that predicts economic *outcomes* but does not describe the mental *processes* that produce them; the classic defense of this view is Friedman 1953.)

Consider an example with cash, something easy to assign numbers to. If you were homeless and destitute, it would mean a great deal to you if we gave you $100. But if you were a billionaire, $100 means almost nothing. Although in both cases the *face value* of the money is identical, the internal *utility* derived from it is very different. Or consider an example without money. You are eating at a Mexican restaurant with friends and the server offers to add cilantro to your dishes. One friend loves cilantro; another has the genetic quirk that makes cilantro taste like soap. Although the physical object, cilantro, is the same for both, the internal utility from adding it to their food is very different.

To bring this together, consider again choosing between $100 for sure and a 50% shot of $250. These need to be converted to a mental currency of utility, which we will call "utils." If you value $100 at 10 utils, then if you choose the sure thing, you get 10 utils. If you value $250 at 16 utils, then if you choose the gamble, you can expect 8 utils (= 16 * 50%). With these utilities, you should prefer the sure thing over the gamble (expected 10 versus 8 utils)—even though the face values of the options predict the opposite (expected $100 versus $125). By supposing an internal currency of utility, a psychological version of expected utility theory can describe why people do not just go for the highest dollar amount. Notice that there is not a straightforward connection between face value and utility: The face value of $250 is 150% larger than $100. But 16 utils is only 60% larger than 10 utils.

Although expected utility theory captures a fair bit about decision making, it also leaves some anomalies unexplained. This is where Kahneman and Tversky stepped in. They developed a theory, called *prospect theory*, that did a better job of capturing how people make decisions involving risk. ("Prospect" was their technical term for the outcome of a gamble.) Prospect theory separates itself from past theories because it assumes that people make decisions with respect to some *reference point*, most commonly the status quo. Where you are right now is, psychologically speaking, special. How much will you *gain* or *lose*? Notions of gain or loss are, of course, very intuitive, but they require a reference point to make sense.

Expected utility theory just evaluates the final outcomes, like how much money you end up with after making a decision. Prospect theory says that people evaluate outcomes relative to where they start; after making a decision, you care not just about how much money you end up with, but whether it's more or less money than you started with. Prospect theory also assumes that *losses loom psychologically larger than gains*. Gaining $100 is nice. Losing $100 is not just the mirror image of the gain, but worse.

We have gone on at length to illustrate that even decisions are quite complex. Researchers are still trying to develop a full theory for how the mind handles decisions, and decisions only involve a single person making a choice. Games, in the technical sense used by social scientists, are what happens when *multiple people's choices all mutually determine the outcome*. In decision-land, although you as an individual cannot control the role of a die or the flip of a coin, you can completely control whether you choose the sure thing or the gamble. With games, even this level of control is gone. What happens to you depends on what other people do, meaning games are *strategic*. Because of this, understanding what people should do and what they actually do in games is even more difficult than understanding decisions. (From this point on, if we use the word "decision," we'll be using it in its everyday sense. So don't worry that we're contradicting ourselves if we say people made a decision while playing a game.)

The complexity of games is why economists and mathematicians devoted a whole branch of study to games, called *game theory*. A touchstone book was the monograph *The Theory of Games and Economic Behavior* by John von Neumann and Oskar Morgenstern (1944). von Neumann was a true polymath. He worked on the Manhattan Project, he helped found computer science, and he has a computer architecture named in his honor (it's probably what your computer is running on right now). He would probably have won a Nobel Prize had he not died young. As crazy as it sounds, expected utility theory, one of the most widely used theories in the history of social science, was developed by von Neumann and Morgenstern as a mere steppingstone for crafting their theory of games. In fact, because it is so much simpler to use than theories like prospect theory, modern game theory often still uses expected utility theory. Let's now look at common games that reveal some features that will be important in the rest of this book.

The Prisoners' Dilemma

By far the most used game in all the behavioral sciences is the prisoners' dilemma. To read every scientific paper about it would take several lifetimes. This game is often used to capture positive-sum trade and exchange

TABLE 1. Possible prison sentences for each set of choices in the prisoners' dilemma

	Crook 2: Stays Silent	**Crook 2**: Rats out Friend
Crook 1: Stays Silent	Each gets 1 year	Crook 1 gets 10 years Crook 2 gets 0 years
Crook 1: Rats out Friend	Crook 1 gets 0 years Crook 2 gets 10 years	Each gets 4 years

(like buying in a store or bartering in a market), cases where there is a possibility for mutual benefit but also some conflict (for instance, I'd rather get my groceries *and* still keep my money). Its genesis is less positive, however, which is why it's called the prisoners' dilemma. The original developers ask us to imagine that the police have arrested two suspected bank robbers. Each person is in a separate interrogation room. The evidence against them is thin; as it stands, each suspect would face only a minor sentence for their crime. But the detectives are willing to give a deal to either suspect if they rat out their compatriot. The stool pigeon would go free; his silent friend would sit behind bars for a stretch. If both rat each other out, however, then they will receive moderate sentences. The table below illustrates this by showing all four possible sets of sentences. If you were one of the crooks alone in your room, what would you do?

In predicting what people would do, let's imagine that the *only* thing the suspects care about is their *own* prison sentence. In particular, we'll assume that years in prison are identical with each person's utility. (We'll see examples later where utility goes beyond personal outcomes.) First, look at the outcomes where both players make the same choice. If both stay silent, that's really good: just a year in prison for both criminals. If both rat each other out, it's four times as bad. So, you might think both should stay silent. Turns out, it's not so simple.

Take the perspective of Crook 1. You're alone in an interrogation room wondering what your buddy is going to do. To start, you consider what to do if you think your friend is going to rat you out. On this assumption, you should rat him out, too. Four years in prison is not great for you, but it's better than 10. Next, you consider what to do if your friend stays silent. You should still rat him out (no honor among thieves). Going free is better than a year in the clink. As it turns out, it does not matter what your friend does: You should always rat him out. The solution to the prisoners' dilemma is easy in this respect. It can be solved with the basic idea of *domination*: A particular choice dominates the other choices available to you if

TABLE 2. Payoffs for each set of choices in the prisoners' dilemma

	Person 2 Cooperates	**Person 2** Defects
Person 1 Cooperates	Each gets $7	Person 1 gets $0 Person 2 gets $9
Person 1 Defects	Person 1 gets $9 Person 2 gets $0	Each gets $2

it's best for you regardless of what the other person does. In this case, ratting your partner out always dominates staying silent.

The prisoners' dilemma is a hypothetical, but thousands of researchers have turned this hypothetical into a game that can be played in the lab. What do actual people do when confronted with this game? Typical experiments put a more positive spin on it: Instead of trying to avoid a cost (i.e., going to prison), people are trying to earn something like money. This maps on to real-world situations like economic trade. Say I have a bag of potatoes and you a cooler of fish. I'd prefer the fish to my potatoes, and you'd prefer the potatoes to your fish, so we're both better off if we trade. But I would be even better off if I get your fish *and* keep my potatoes. So, there's a tension between mutually beneficial exchange and successfully cheating each other.

The following table shows an example of a laboratory prisoners' dilemma with dollar payoffs. (To map back to trade, you can imagine that each player values the other person's goods at $7 and their own at only $2.) Despite the positive spin, the prediction is the same: No matter what the other person is planning to do, it's best for you to defect and keep your potatoes.

When real people play games like this, even when the play is completely anonymous, the players cooperate with each other way more than would be expected by our analysis. This is in part because people care about more than their own payoffs. Another reason is that the game is often played multiple times, with the same pair of people engaging in multiple "rounds" of the same game. When the prisoners' dilemma was originally devised as a thought experiment, the researchers immediately had real people play it in an iterated format and found that they cooperated about 70% of the time (Poundstone 1992). Our analysis of what to do assumed that the game is played just once. If it's played back and forth, much more cooperation is possible (Andreoni and Miller 1993). In our own research, using mathematical models, we have shown that even if the game seems to be a one-off encounter, people should often treat it as if it's repeated and therefore they should cooperate (Delton et al. 2011a, 2011b; Delton and Krasnow 2014; Zimmermann and Efferson 2017; Zefferman 2014; McNally and Tanner 2011).

Gone Hunting

Let's look at a different game, one whose solution is a bit more complicated. This game, the stag-hunt game, is based on a thought experiment created by the eighteenth-century Genevan philosopher Jean-Jacques Rousseau. Imagine you and a friend are hunter-gatherers and there are two spots where you can go hunting. In one spot, there are stags to hunt, but it takes both of you to bring one down. If you show up alone, you get nothing. In the other spot there are hares, and you can catch a hare whether or not your friend comes; hares are so small they only feed one person. The spots are far apart, so once you go to one spot, that's it for the day. The key is that *splitting* a stag is way better than having an *entire* hare for yourself. Unfortunately, because you lack phones, you cannot just ask your friend where he's headed. You'll have to guess and hope for the best. The possibilities are shown in the tables below, first in terms of stags and hares (the real-world game) and then a version with money (a possible laboratory game). What would you do?

Again, let's assume all you care about are your own outcomes, and let's look at the game in terms of money instead of stags and hares. If you are Hunter 1, what should you do if Hunter 2 plays stag? You should play stag also because $7 is better than $2. What should you do if Hunter 2 plays hare? You should also play hare because $2 is better than $0. It's best to

TABLE 3. Payouts in food and in money for each pair of choices in the stag-hunt game

	Hunter 2: Stag Location	**Hunter 2**: Hare Location
Hunter 1: Stag Location	Each gets half a stag	Hunter 1 gets nothing Hunter 2 gets a hare
Hunter 1: Hare Location	Hunter 1 gets a hare Hunter 2 gets nothing	Each gets a hare

	Hunter 2: Stag Location	**Hunter 2**: Hare Location
Hunter 1: Stag Location	Each gets $7	Hunter 1 gets $0 Hunter 2 gets $2
Hunter 1: Hare Location	Hunter 1 gets $2 Hunter 2 gets $0	Each gets $2

match whatever your friend is doing. This is very different from the prisoners' dilemma, where defecting dominated cooperating regardless of the other person's choice.

A way to think about games like the stag-hunt game was discovered by Nobel laureate John Nash. Nash became famous to the public from the biopic *A Beautiful Mind*, which showcased his successes as a mathematician and his struggles with mental illness. Nash developed an idea now called the *Nash equilibrium*. In a game, a Nash equilibrium is a set of choices that are the best responses to the others' choices—every single player should think that, given what the other players are doing, the focal player is making the best choice possible. Another way of saying this is that at an equilibrium, no player would want to *unilaterally* change their choice. If you and your friend were both going to the stag spot, there's no reason for you to unilaterally change to the hare location. And if you're both going to the hare location, you would not want to unilaterally change to the stag location—you cannot catch the stag alone. There are two Nash equilibria in this game, pairs of choices that, once arrived at, each player would want to stick with. (You might notice in this definition that mutual defection in the prisoners' dilemma is a Nash equilibrium. However, the idea of a Nash equilibrium is not required to see why players would end up at that pair of decisions; you only need the simpler idea of domination.)

Which equilibrium should players choose? One proposal for how to select between multiple equilibria is to assume players will pick the "payoff-dominant" one, that is, the equilibrium that is better for everyone. In this game, both players are better off if they both play stag than if they both play hare. A different proposal is that players may be concerned with the relative risks of trying to play each equilibrium. If you try to play stag but your friend doesn't, you end up with nothing. But if you play hare, you end up with $2 no matter what your friend does. So, the less risky equilibrium is perhaps the way to go. Evidence suggests that both considerations are on real players' minds.

The stag-hunt game is part of a broader class of games called *coordination games*, which have multiple possible equilibria. This is relevant to our book because, as we'll see later in this chapter, the primary game we focus on—the disaster game—is a coordination game.

What's in a Game?

Now that we've seen a few games, let's step back and consider the process of turning a real-world game into a laboratory experiment. The actual prisoners' dilemma is hypothetical: No research scientist ever captured two alleged bank robbers and put them in separate interrogation rooms. How can scientists study in the lab the strategy at play between these two criminals? *How can scientists create economic games?* The way we think about creating a game draws heavily from Nobel laureates Vernon Smith and Elinor Ostrom (Ostrom, Gardner, and Walker 1994; Smith 1982, 1994).

Ostrom and her collaborators encourage scientists to answer several questions about their research when designing laboratory games. To start, you need to know something about the people you're studying. This means thinking about their traits, goals, resources, knowledge, and more. For instance, what kinds of roles or positions do these people occupy? Some people may be leaders and some followers—your governor is a leader and the people in your state are followers. Or people may have different but nonhierarchical roles, as in a business where some people are responsible for technical development, others for accounting, and still others for human resources. Or everyone may have identical roles, like students in a class. In the prisoners' dilemma, both suspects are identical: They know that both face the same situation and they both want, all else equal, to spend less time in jail. The only difference between the two is that each suspect only knows their own mind; they do not know what the other person will do.

To make a game, you also need to know what actions each person can take. In the prisoners' dilemma the actions are to keep silent or to snitch. In a committee, reasonable actions may be voting yea or nay on a proposal or making one's own proposal. When students work on a group project, the decision may be how much time or effort to invest in the group project versus other assignments.

Another task for the researcher is to identify how everyone's actions cause particular outcomes, and what kinds of costs and benefits these outcomes create. In the prisoners' dilemma, you cannot really separate outcomes from costs and benefits: If both make the choice of staying silent, then the *outcome* is that each gets one year in prison; we could just as well say the *cost* of both staying silent is one year in prison. In other cases, there is a greater distinction between outcomes versus costs and benefits. A work team may produce a product, like a marketing campaign. The same campaign could be produced (more or less) by different combinations of effort from team members. Maybe Jill stays nights and weekends while Sam spends his days surfing the web. Or perhaps the reverse is true. Either way, the team as whole might still produce a good campaign. Yet to the extent that their boss can monitor individual performance, Jill might get a raise and Sam might get fired.

Ostrom stresses that although this analysis of the situation can be connected to game theory, it's not the same as game theory. Game theory is a family of mathematical models that researchers can use to predict or understand behavior in some strategic arena. But correctly describing the situation is logically prior to connecting it to game theory. To see this, notice that we can verbally describe the prisoners' dilemma without any math, and moving it to formal game theory requires additional assumptions. For instance, in standard game theory, we assume that agents' preferences and choices can be described by expected utility theory (a simplifying assumption that, as we have seen, is not always correct). Standard game theory also makes assumptions about people's ability to think strategically, such as their ability to predict how other people will respond to their own choices. When you play chess, you must anticipate how your opponent will move their pieces in response to your moves. Standard game theory assumes a basically infinite ability to think through possible moves and countermoves. Real people fall short of this ideal (see Camerer 2011).

So far in this section we have talked about understanding strategic arenas that do or could exist in the world and connecting those to math models. But what about *creating strategic worlds in the lab*? Vernon Smith has written extensively on this. The goal is to create a laboratory task where we as researchers fully understand the problems people face, their possible choices, and their material incentives. Smith created laboratory markets where he could test how people trade goods and commodities and how different rules governing the markets affect trade. This allowed him to compare what real people do with theories from economics (e.g., Smith 1982).

So, one feature of designing a good game is careful selection of the *rules* or *institutions* that govern players' outcomes. Smith has studied markets such as auctions. In the real world, some auctions are English auctions, the kind you see in movies. The auctioneer announces a starting bid, for example for a newly discovered Monet, and the members of the audience each signal their own, progressively higher bids. When the auctioneer decides no one else will bid—going once, going twice, sold!—the highest bidder pays their bid and gets the painting. Other auctions can have different rules. In a "first-price sealed bid" auction, everyone submits a private bid at the same time. Unlike the English auction, bidders have no knowledge of others' bids and no way to adjust their own bids. Whoever bids the most pays that amount for the painting. Because the rules are different, there's the possibility that they will raise different amounts of money for the same painting, or be better or worse at revealing how much the bidders truly value the painting. By creating laboratory versions of these auctions, researchers can see how real people behave in each auction.

Games can also be used to study political institutions like voting rules (Morton and Williams 2010). In research led by Peter DeScioli, we used game experiments to study how people make decisions that are particularly bad for a minority group (DeScioli et al. 2018). A government planner might determine that putting a highway through some area of town will stimulate the local economy, lifting most people's incomes. However, the people living right next to the highway will face more noise and traffic, and a depressed economy as businesses are demolished to make way for the road. One policy would be to build the road, which benefits the majority at the expense of the minority. Another would be to shelve the road and leave people as they are. We created a game that captured this kind of problem. We asked people to decide between policies like these using simple majority voting—an institution often used in the real world. We then compared this to an institution that does not exist in the real world. Here, one player was randomly selected, and their decision alone determined the outcome. We assumed players in the minority would always avoid the policy that hurts them. The question is: What do majority players do? Some researchers have suggested that voting creates a community spirit. If so, majority players should be more likely to *vote* against the majority-favoring policy than when they are randomly selected to decide. When we ran this game, we found that many majority players voted *against* hurting the minority. However, the institution did not matter: The player selected to be decisive decided against hurting the minority about as often as voters did.

In the lab, like in the real world, the rules and institutions must lead to tangible outcomes and payoffs for the players. What should those be? In Smith's market experiments, one option would be to bring in real objects, like coffee mugs or T-shirts. But people could bring with them all sorts of weird ideas and preferences regarding coffee or cotton. Maybe you have a favorite coffee mug at home, so you have no interest in bidding on a new mug. Or maybe you broke your favorite mug on your way to the lab, so you're desperate for a new way to hold your drink. As researchers, we don't know all of the different preferences people might have over objects like mugs, so market experiments using mugs leave the researcher in the dark about why some people loved them and some hated them.

Knowing this, Smith instead had people in his experiments trade abstract laboratory commodities—basically, pieces of paper that could be redeemed for money. Smith and his collaborators would tell each player how much the commodity was worth to that player personally. For instance, an experimental *buyer* may be told that if they are holding the commodity at the end of the experiment, they will receive $10. Thus, if they can buy it for less than $10, they make money. But an experimental *seller* may be told that that if they are holding the same commodity when the experiment ends, they will receive $5. If they can sell it for more than $5, they make money. Thus, the same outcome—holding the commodity—has different costs and benefits for different people. Smith created laboratory incentives that match real-world markets: In general, buyers value a particular commodity more than sellers do; that's why they are buyers and sellers, after all. And because he created the values, Smith completely understands them—unlike preferences for mugs or shirts.

Although there are many possible ways you could reward people, like cookies, juice, or pats on the back, Smith argues that money is particularly useful. There are practical benefits because it's easy to count, it doesn't perish, everyone knows what it is, and so on. But in Smith's view, perhaps the best feature is that people do not "satiate" on it. What he means is that when all other features of a choice are held constant, people will always pick the choice that gives them more cash. Imagine you are already in our lab with your hands over a keyboard. We tell you to type "1" for $1 or type "9" for $9. You will of course type "9." More money is always better. This is not true of everything. Eventually you'd get sick of cookies or juice, or you would run out of places to store them. Who wants to get paid with a million cookies? Perhaps if people were choosing between enormous sums like $4.1 and $4.5 billion, they would be indifferent. (We doubt it.) But for any amount realistically used in a lab, more is better. Without this feature, it would be much harder to experimentally control material incentives: Everyone wants more money (again, all else equal), but how many cookies people can eat in one sitting varies wildly.

A second feature Smith emphasizes is that laboratory rewards must be "salient." This means that what you earn must be connected to your choices (and the choices of others). Game researchers usually pay subjects a flat fee just for showing up at the lab, plus additional money that subjects earn from the choices they and others make in the experiment. The show-up fee is not salient in this technical sense; a subject gets it even if she falls asleep at the computer. (We've had to wake up a few people in our experiments over the years. It's awkward for everyone.) The money from choices, however, is salient. One choice might earn you $10 and a different choice might leave you broke—depending, of course, on what other players do.

When rewards are salient, the researcher can match the lab rewards with the incentives presented by real-world problems. Remember our prisoners' dilemma: First, we described the game verbally, laying out which outcomes were better than others and their rough magnitudes. The suspects wanted to avoid jail time. They got the least jail time for all staying silent and the most jail time if they both snitched. Second, we translated this description into dollar amounts that (to us) seemed to match the verbal description. This is as much art as science. In our telling of the prisoners' dilemma, the difference between both suspects ratting each other out and both staying silent is represented by $2 versus $7. Is that the correct difference in magnitude? Who knows! There's a leap of faith here; intuition is often the final guide for exact numbers.

Third, Smith recommends that the rewards be "dominant." This means that the rewards the researcher wants to study should be the only costs and benefits that matter. For instance, imagine that a researcher creates a game that is exceedingly difficult for players to understand. The instructions are written for other scientists and not lay people, the rules are explained using complex derivatives and differential equations, and the instructions go on for pages. For players, the burden of understanding the game may outweigh any conceivable benefits from choosing wisely. Unless the researcher wants to know whether complicated instructions cause people to trade less, this is a bad design. The burden of understanding is now part of the game, even though it's not what the researcher wants to study. That is, players aren't just deciding what decisions to make based on the rules the researcher cares about, they are also deciding whether they should even take the time to read the rules. All else equal, games should be transparent to the people playing them.

Do You Care?

In typical markets, like an exchange floor brimming with stock traders, the traders do not care about what happens to one another. They want to buy stocks as cheaply as possible and sell for as much as possible. If you try to create this in the lab, often with university students, one worry is that your student players will want to be nice to one another even at personal expense. They might look around, see people they will later pass in the halls or people they take classes with, and feel solidarity or warmth toward them. Maybe your players will *care* what happens to the other players.

To skirt this problem, Smith's final recommendation is that rewards should be "private." This means that each player in the experiment knows how much they will earn from an experimental commodity, but they have no idea what the other player will receive for the same commodity. Without this information, each player cannot purposefully help or hurt others—their only option is to earn as much as they can for themselves. Plus, this captures the essence of trade and exchange: The reason to buy something is that you value what you're buying more than you do the money in your pocket. Giving people different valuations for the same commodity does this in a way that is transparent to the researcher.

Anonymous market interactions, however, are not the only types of interactions we might be interested in studying. Sometimes we *do* care what happens to other people and have the information to affect others' outcomes. So, we need games that can model this. By far the most popular game for doing so is the *dictator game*. This game is super simple: In the canonical version, two people are paired up through a computer network (this way they do not know each other's identities). One of them is randomly selected as the dictator. The dictator is given a stake of money, say $10, and can divide it any way they like between themselves and their partner. The dictator can keep the entire $10 for themselves, give it all to their partner, or anything in between. That's it. The partner has no say in what happens. In fact, since the partner has no say, this game is not quite a game—there's no strategy involved for the dictator because they can do whatever they want. Instead, the game is taken to reveal "social preferences," generally measured in how much money the dictator gives to the other player. But a social preference for what exactly? Fairness, equality, equity? It's an area of lively debate (see more in our Chapter 4).

Regardless of exactly what the dictator game measures, people are quite generous. Thousands of dictator-game studies have been run and it's impossible to summarize them here. One great study, by a team of anthropologists, took the dictator game to societies throughout the world (Henrich et al. 2005). They mostly played the game with people from small-scale societies, meaning hunter-gatherers, horticulturalists, pastoralists, and the like. They also included a few games with people from mass societies like the US. Despite stupendous variability in the societies studied, the results revealed a shocking uniformity. In principle, dictators could give anything from 0% to 100% of the stake to their partner. In reality, the average amount given in each society ranged within a small window of about 25% to 50% of the stake. The researchers also found that what variability did exist could be predicted by how much the societies were integrated into markets—more integration, more generosity.

The Public Good

Tools like the dictator game are critical for our goals in this book. Many questions concerning climate change involve a willingness to sacrifice one's own material benefits for other people. Are you willing to drive less and bike more to reduce your carbon footprint? Is your country willing to make the transition from fossil fuels to renewable energy to save future generations from climate disaster?

But climate change involves other problems as well. One important problem is that climate change mitigation has the properties of a *public good*. When something is a public good, everyone benefits from its existence, but each individual is best off not paying for it and letting others take care of the problem. Consider the problem of law and order. Everyone benefits from having a functioning, well-trained police force and judiciary. But if you personally withheld your tax money, you'd have more money in your pocket and the police and courts would be unaffected (after all, your own taxes are a tiny drop in a large bucket). So, personally, you would like the best of both worlds: the protection of law and order while keeping your own cash. The problem is that if everyone thinks this way and shirks on their taxes, no one gets law and order. Other common examples of public goods include national defense, roads, and clean water. In all these examples, everyone is better off when these things exist, but each person would be even better off if they let everyone else fund them. As we'll detail more later, mitigating climate change is also a public good because everyone benefits from it, but each individual person's actions, or even the actions of most countries, are not strictly required to solve the problem—so why pay bother paying the costs? (You may have noticed that public goods sound suspiciously like the social dilemmas we mentioned in the last chapter. You're right. Public goods are a subset of social dilemmas.)

Given how important these kinds of problems are, it's not surprising that researchers have developed a game to capture the essence of public goods. It's obvious that a powerful ruler could simply use force to create public goods, should they desire to. Researchers have been more interested in understanding whether it's at all possible for public goods to be created from the ground up by equal people. Although exact details vary, a typical public goods game might go like this. Four players work as a group. They interact anonymously over a computer network and will never know one another's identities. Each is given $10 to begin; see Figure 1. Each person independently decides how much to contribute to a joint pot; any money a

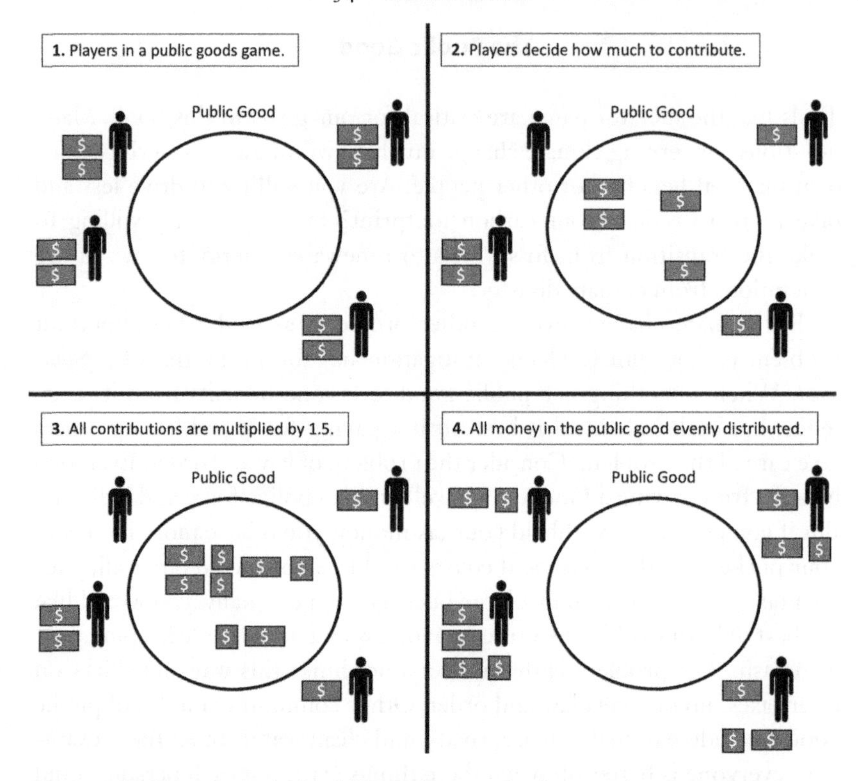

Fig. 1. Each phase of a typical public goods game. In phase 1, players each start with a personal pot of money. In phase 2, they decide how much of their own money to give to a group pot. In phase 3, the experimenter multiplies whatever was contributed; in this example, contributions are multiplied by 1.5. In phase 4, the multiplied money is divided evenly to every person. The tension of the game arises because the group as a whole does best if everyone contributes, yet individual people do best if they share in others' contributions without contributing themselves.

player does not contribute is theirs to keep. The experimenter then takes any money in the joint pot and multiplies it by 1.5. This captures how cooperation is synergistic, such as how everyone is better off when law and order is created. This new, bigger group pot is then divided equally among all four players, regardless of how much each individually contributed. (Any money players held on to is left as is.)

Consider the material incentives for the group and for each person. If no one contributes anything to the joint pot, everyone leaves the study with $10. And if everyone contributes everything to the joint pot, everyone leaves the study with $15, a 50% return on a $10 investment.

So, is it a no-brainer to contribute to the common pot? No. Each person is even better off if everyone else contributes but the individual holds onto their cash. Consider what happens if three people contribute everything and the final person contributes nothing. The three contributors each earn $11.25 (= 3 * $10 * 1.5 / 4). This is barely better than not contributing at all. And the one noncontributor—called a *free rider*—earns a massive $21.25 (= 10 + 3 * $10 * 1.5 / 4); in other words, they keep their original $10 and *also* get the shared benefits created by their cooperative group members. This maps onto real-world public goods: Everyone benefits if roads or clean water are created but the individual incentives are to shirk. Like the prisoners' dilemma, the equilibrium is for everyone to defect and contribute nothing to the public good.

What do real players do? As with the dictator game, so many public goods experiments have been run it's impossible to summarize them all here (Ledyard 1995; Chaudhuri 2011; Balliet 2010; Balliet and Van Lange 2013). A key takeaway is that most people contribute *something* to the public good, but rarely do they contribute everything.

Researchers have also layered additional institutions on top of the basic game to see if this improves cooperation. For instance, if players can communicate, they tend to be more cooperative (Sally 1995). Partly, communication allows players to generate a plan of how best to play—they can figure out how to all walk away with as much money as possible and tell each other how much they plan to contribute. And it also seems to create solidarity among the group regardless of any specific plan, reducing the number of people who free ride. Another institution that makes cooperation more likely is the possibility of punishing people who do not cooperate (Fehr and Gächter 2000; Yamagishi 1986). Sometimes this is "peer punishment," where each player is given the option to punish anyone they choose (and, usually, they punish people who refused to contribute to the public good [Fehr and Gächter 2002]). As with the prisoners' dilemma, researchers studying public goods games often have the same group play a series of "rounds" of the game in a row. The typical finding is that contributions start off reasonably high but decline over time, seemingly because contributors see others are not contributing as much as them and lower their contributions in response, which creates a vicious cycle until few people are contributing (Camerer 2011).

The public goods we've been discussing so far are, in technical terms, "linear" public goods games. This is because every dollar contributed creates the same amount of collective benefits. The first $1 is multiplied by 1.5, and creates a collective benefit of $1.5. The second $1 contributed also creates a $1.5 collective benefit, just like the thousandth $1 contributed creates $1.5 in collective benefits. No matter how many dollars are contributed, the next $1 contributed creates $1.5 in benefits; collective benefits scale linearly with contributions.

A variant of the public goods game that will be especially important for the rest of this book is the *threshold* public goods game. The threshold game gives players a target they must meet with their contributions before they can earn any extra money. For instance, imagine four players each starting with $10. If the group as a whole contributes at least $20 to the public good, then everyone will get a personal bonus of $10. If all four players contribute $5 each, they exactly meet the threshold, each gets the bonus, and they leave the experiment with $15 (= the $10 bonus plus their remaining $5 personal fund). Typically, whatever money is contributed to the threshold disappears regardless of whether the threshold is met.

Introducing the threshold is a simple change to the rules that has drastic effects on the properties of the game. Whereas the original public goods game is like the prisoners' dilemma because everyone's material incentive is to defect, the threshold game is like the stag hunt because it's a coordination game. Recall the idea of a Nash equilibrium: At an equilibrium, no one would want to *unilaterally deviate*. Let's see how this plays out in the threshold game. Imagine that everyone in the group is contributing exactly $5. Would one player want to switch and contribute nothing? Nope. If this one player switched, then no one, *including this player*, gets any bonus. And receiving the bonus for a net $15 is better than only getting their original $10. What if everyone is contributing $0—should one player unilaterally start contributing? Again, no. Even if a single player gave all their money, they alone cannot meet the threshold, so there is no point—they're just throwing money in the trash. Right away, we can see there are at least two equilibria in this game: everyone contributes exactly $5 or everyone contributes nothing.

It turns out the threshold game has a huge number of equilibria. The first is that no one contributes anything at all. The rest of the equilibria are *every possible combination of contributions that exactly meet the threshold.* Consider a case where one player contributes $7, one $8, one $5, and the final player $0, adding up exactly to the threshold of $20. Would anyone want to unilaterally deviate? No. If anyone raised their contribution, it's pointless—they're already going to get the bonus, so contributing more is a waste. And if any of the players decrease their contribution, that's really bad—removing any dollar means no one gets a bonus, including the player who "saved" a dollar.

So, how should players settle on an equilibrium, especially when they can't communicate with each other? When the players are otherwise identical, the most obvious way to coordinate is where we started: Everyone contributes the same amount and just enough to collectively meet the threshold. Compared to some random combination of contributions, like the example in the previous paragraph, a scheme of equal shares is far more intuitively compelling (see Schelling 1960; Fiske 1992). On the other hand, there is still a tension between group- and self-interest: After all, any one player benefits if they dig their heels in and let everyone else meet the threshold.

When real people play social dilemmas, they are more likely to contribute when they see others doing so as well (Fischbacher, Gächter, and Fehr 2001; Kocher et al. 2008). More broadly, contributions in the threshold game are usually higher than the standard public goods game (Jordan, Jordan, and Rand 2017). This makes sense because defecting no longer dominates every other strategy. If there is a threshold for success, success is easier. This is important for thinking about climate change because many of the most worrisome problems involve thresholds, points of no return like tipping points (see Chapter 3).

From Goods to Bads

There's a key difference, though, between the games we've just covered and ones that capture the strategy of preventing disaster. In public goods games, whether the linear version or the threshold version, people are working together to create a shared *benefit*. If we all contribute, we're all better off than we were before. But when we prevent disaster, we're working together to prevent a *bad* thing from happening. An economist might say that in a certain logical sense creating a public good and preventing a public bad are the same (Cornes and Sandler 1996); either way we are working together to end up better off than we would be if we didn't work together. Indeed, the most common example of a public "good" in textbooks is national defense, which is really the prevention of a bad—foreign invasion. But as we saw, Tversky and Kahneman have shown that real people judge outcomes relative to a reference point, so it probably matters whether a game is framed in terms of creating goods or preventing bads.

At long last, this leads us to describe the workhorse game that we will use in much of this book: the *disaster game* (Milinski et al. 2008). (To avoid too much repetition, we defer extensive discussion until the next chapter

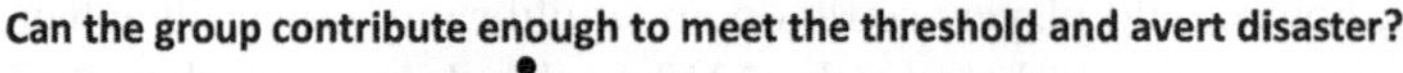

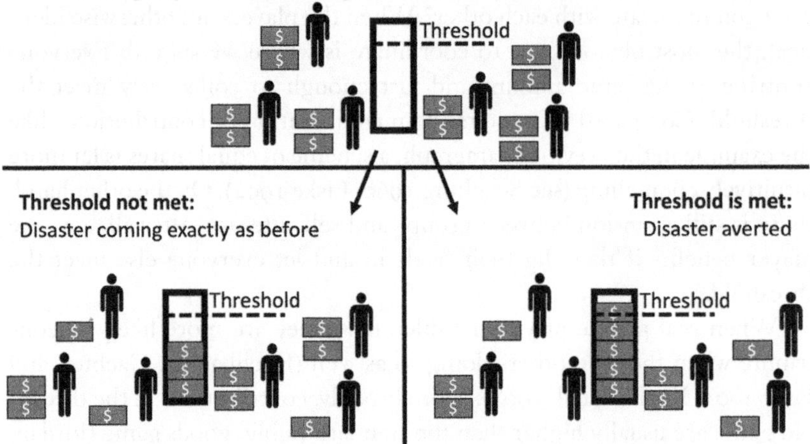

Fig. 2. The disaster game. Everyone starts with personal money (top panel). They each decide how much to contribute to a group threshold. If the combined contributions are less than the threshold, everyone loses their remaining money (bottom left panel). If the combined contributions are more than or equal to the threshold, everyone keeps their remaining money. (Contributions are lost regardless.)

and only describe the most basic features here.) The disaster game involves a straightforward change from the threshold public goods game: Instead of earning money by meeting the threshold, groups try to meet the threshold *to prevent their money from being destroyed*. For example, a group of 4 players might each start with \$10. If the group collectively contributes \$20 or more, then everyone keeps whatever money they did not contribute; see Figure 2. If the group contributes less than \$20, then they lose everything.

Because the disaster game involves a threshold, it's a coordination game. For instance, if everyone contributes \$5 in this example, then they have met the threshold and all keep their remaining \$5. No one would unilaterally want to deviate. If anyone kicked in a dollar more, it would be pointless since they've already averted disaster. If someone contributed even one dollar less, the threshold would not be met. The shirker doesn't benefit from keeping the extra dollar: Since they failed to meet the threshold, they all lose everything—including the shirker. As before, nearly any set of contributions that adds up to exactly \$20 is a Nash equilibrium that no one would deviate from. Also as before, it is a Nash equilibrium for no one to contribute anything. No player can meet the threshold on their own (\$20 threshold > \$10 stake). So, if no one else is contributing, it would be pointless for a lone player to do so. Players in the disaster game often converge on equal contributions (Milinski et al. 2008; Barrett and Dannenberg 2012, 2014).

To recap the disaster game, players begin with a pot of money that will disappear if they cannot prevent disaster. The group is given a threshold, which is a dollar amount. If they contribute less than the threshold, disaster strikes and they lose everything. If they contribute at least as much as the threshold, they avert disaster and keep their remaining funds. They do not get back the money that they contribute to the threshold, no matter what happens.

The Utility of Economic Games

Any game is only useful insofar as it successfully captures what a researcher wants to understand. The strategic problem being studied could be one that currently exists, or it could be a hypothetical, future problem. The disaster game can be applied to both types of problems. Critically, many scientists believe that climate tipping points are a real problem facing the world. Tipping points are changes in the climate with sudden, sharp discontinuities that lead to catastrophic loss (Lenton et al. 2008). Once we cross a tipping point in the climate system, it's difficult or impossible to change the world back to how it was before we passed the tipping point. As we discuss in detail in the next chapter, the disaster game is best thought of as modeling mitigation to prevent reaching these tipping points. This is because the game itself has a sharp discontinuity: If players meet the threshold, then they avert disaster. If they contribute any amount less than the threshold, the disaster still happens, and if they contribute above the threshold there are no additional benefits to anyone.

Key to our book is that games can model strategic problems that, at the moment, are purely hypothetical. Looming tipping points already exist; the disaster game can model this real problem. But many issues involving climate disaster are ones humanity will only face in the future. For instance, right now no technology exists that can successfully modify radiation from the sun, at least none that can be rolled out on a wide scale (see Chapters 3 and 6). Still, it would be useful to study how people might react once we have the option to use solar geoengineering. One of the beauties of games is that we can create strategic problems in the lab even if they have never existed in the real world (Morton and Williams 2010).

A second benefit of games is that we can *randomly assign* players to different conditions. If you've never taken a course in research methods or experimental design, you might not realize just how revolutionary the idea of random assignment is (Shadish et al. 2002). Science works in part by making comparisons. If we drop a five-pound weight in a vacuum, will it fall faster than a one-ounce weight? It turns out it will not; gravity accelerates all objects the same. This fact can be learned without recourse to a randomized experiment. But does a new pain reliever work better than the old standard? Without randomization, it would be hard to tell if the new pill is effective. Maybe the people taking the new pill already had less overall pain for some reason. By randomizing people to take either the new pill or a placebo pill, we ensure that all possible preexisting differences between people—like their pain tolerance or existing pain—will be randomly shuffled among the two different groups. Yes, individual *people* will differ, but the *groups* will not on average differ before getting whichever pill they are assigned. And, of course, this extends beyond pills. By randomizing in games, we ensure that players with different levels of generosity or different abilities to think strategically are spread evenly across different conditions. We can then see if our experimental manipulation drives differences between the groups.

A final benefit of experiments builds on the previous two. Take for example the case of geoengineering. We might be interested in whether giving people the opportunity to invest in geoengineering increases their ability to stop climate change (see Chapter 6). But it would be both impractical and unethical to randomly assign some countries to be allowed to use geoengineering while stopping others from doing the same. In the lab, however, we can create an analogue of geoengineering and randomly assign people to have or not have it, all without the same concerns. Practically, it is easy to randomly assign people to experimental conditions (computer programs for research have this function built in). And, ethically, we aren't actually having people invest in geoengineering, but instead are testing how they respond to a game that models those strategic features. Players in our games are playing for $20, rather than the fate of the world in the face of climate change.

Questions?

What would it mean for the disaster game and similar games if it turned out that they do not map onto the strategic problems of climate change? For instance, if tipping points turned out to be irrelevant to climate change, then the disaster game, with its threshold logic, would be a poor guide to understanding how real people might make decisions about climate change. Whether our games are useful for understanding climate change depends on at least three things. First, climate scientists must, with reasonable accuracy, correctly describe climate problems. Second, people such as politicians, engineers, and economists must provide a reasonably complete picture of the alternatives available to the world to deal with these problems. Third, we must design our games to match these problems and potential solutions.

None of these steps are trivial. The one we can control, our games, could go wrong in several ways. Obviously, we could simply do a bad job of capturing the problem: If climate scientists said tipping points were not the problem, but we ploughed ahead using the disaster game with a threshold, our games would not tell us much about real-world climate politics. Although this is a problem, it does not *necessarily* render the disaster game useless. It might not capture climate politics, but there are many other disasters with thresholds. Building a levee to stop a flood certainly has a threshold. If the levee is too low, the town floods. Pandemics also have clear thresholds. If the average infected person passes on their infection to *more* than one new person, a disease spreads exponentially (until herd immunity is reached); if an infected person infects *less* than one new person, the disease dies out or at least circulates at low levels. Even if it turns out that the disaster game does not model climate problems, we are still learning how people think about solving other disasters. (At least, that's what we tell ourselves.)

Games can also go wrong in less obvious ways. Games require simplification so that real people can understand and play them. But if we make the game too simple, it might miss important features of the strategic problem. In the basic version of the disaster game, the size of the threshold is fixed and known with certainty. This makes the game simple to communicate, but it comes at a price: No real disaster has such a perfectly known threshold. Is this a problem for the disaster game? As we show next chapter, there are some qualifications to consider when using a perfectly known threshold, but it still allows us to learn how people behave in the face of disaster.

Games also leave out important parts of real politics in other ways. It's not easy in games to create analogs of long-term group affiliations like partisanship. Sure, an experimenter can divide people into groups based on whether they are already Democrats or Republicans. But this is not an experimental manipulation; there's no random assignment. So, if Democrats and Republicans behave differently in the game, it's difficult to know whether it's their partisanship itself, or something that goes along with partisanship, that causes the difference. It's true that researchers can assign people to novel experimental groups (for example, by flipping a coin to assign people to a red or a blue group). Studies like this capture some aspects of groupishness, such as showing that subjects immediately like their group members more despite not knowing them or ever interacting with them (Tajfel and Turner 1979). Nonetheless, no one would think these laboratory groups capture the totality of partisanship, ideology, or other long-running features of people's political or social alliances. These are important to climate politics but difficult to create in games.

Readers might also wonder whether games capture how a person would behave in the real world. Would a person who contributes a lot while playing the disaster game also contribute a lot in a real-world disaster? Although this may be an interesting question, it misses the point of games, at least the way we use them. Most games are not designed to be "assessment tools" in the way an intelligence test is. Researchers design intelligence tests so that they can use a person's answers on a piece of paper to predict whether that person will succeed in work and life. The goal of these tests is to make a good guess *about a particular person's* outcomes. Or think about assessing major depression. The goal of a depression screening is to determine whether *this particular patient* is, in their real life, suffering from depression. Sometimes games like the dictator game are used for this kind of purpose. Researchers might ask people to play a dictator game and then use their generosity with an anonymous partner as a sort of personality measure, a measure of how generally generous they are.

This is a fair use of games, but it's often not the reason researchers use them and not our goal here. We use games to model important strategic problems and see how people—in general—respond. Whether a *particular* person behaves similarly in a real-world problem that seems to match the game does not tell us much. Out in the real world, the person may have incentives that we are unaware of but that matter for the choices they make. Or there may be additional rules or institutions that constrain their behavior that are beyond those we are interested in. Either way, because the game is not meant to serve as an assessment tool, it's not really relevant whether the *same* person behaves the *same* way when playing our lab game versus behaving in a real-world analog. Instead, the goal of the game is to characterize what people in general will do. When human beings are presented with a strategic problem, can they solve it? And how do different institutions change our ability to solve these problems? With a game we can capture the problem (albeit in a simplified way), we can know everything about it because we created it, and we can see what real people do.

But this leaves open a question that we do think is important: Will the *type* of person who usually participates in games be a good stand-in for all people or for the particular people making climate decisions? As we mentioned before, anthropologists have taken simple games like the dictator game all over the world and found shocking levels of convergence across widely different cultures (Henrich et al. 2005). There are of course quantitative differences across societies, but people everywhere play these games similarly. In our own research, we often collect data from two different sets of people: internet samples of diverse American adults (who participate wherever they happen to be) and samples of undergraduates at our university (who participate in the more controlled setting of our physical lab). We have yet to see a meaningful difference between these samples.

Still, major climate decisions will not be made, at least not directly, by random people on the internet or students in our lab. Politicians and other elites have the most power in these decisions, and they might differ from citizens on many dimensions—income, education, intelligence, experience, and so on. Nonetheless, do typical people and elites make decisions in the same way? It seems like the answer is often yes. Brad LeVeck and colleagues (LeVeck et al. 2014) had 102 elites play a common game of bargaining. These elite players averaged over two decades of experience in high-level negotiations, including stints in top corporate positions and in the US House of Representatives, the Department of State, and the Department of the Treasury. The elites had much more experience with high-stakes bargaining than the average person. Nonetheless, these elites were similar to typical people in that both tended to reject unreasonably low offers made by a bargaining partner. Still, there were differences: The elites were somewhat more likely than a typical sample to reject low offers and showed more strategic sophistication. Although elites are not literally the same as typical citizens (no one would have expected them to be), both nonetheless played the game similarly.

Another relevant study comes from Lior Sheffer and colleagues (Sheffer et al. 2018). They did not study games but other difficult problems, like those Kahneman and Tversky used to claim that people are biased or less than optimal when they make decisions. Sheffer and colleagues collected data from normal people and from nearly 400 national legislators in Canada, Belgium, and Israel. Across many different types of decisions, they found similar choices in politicians and everyday people, including choices that seem anomalous according to math or logic. For instance, one task was drawn directly from Kahneman and Tversky. In it, people had to decide how to deal with a (hypothetical) pandemic by choosing between two policies. There are 600 lives at stake. One policy will definitely save just 200 of them. The other policy has a 1/3 chance of saving everyone and 2/3 chance of saving no one. This is the description that half the subjects got—a description in terms of *gains* (i.e., lives saved). The other half of the subjects had the same objective facts described to them, but in the language of *loss* (e.g., under the first policy, 400 people will definitely die). Much research has shown that when this problem is framed in terms of *gains*, subjects usually want the *safe* option (i.e., saving 200 people); however, when it is framed in terms of *losses*, they shift to preferring the *risky* option (i.e., a 1/3 chance no one will die). What many researchers find striking about this is that although everyone is presented with the same objective facts, the way it is described changes people's choices. Does this extend to politicians? Yes: When the game was described in terms of gains, about 30–40% of people—typical citizens and politicians alike—chose the risky option; when the game was described in terms of losses, about 70–80% of citizens and politicians chose the risky option.

Although we would not claim that there will never be a meaningful difference between elites and typical game subjects, the available evidence suggests that decisions will be made similarly by both types of people. Further, studying typical people in the disaster game, even if it does not tell us exactly what politicians would do, nonetheless tells us something: It tells us how *those typical people want the game to be played*. This is not trivial. Ultimately, politicians and other policymakers are constrained (at least in democracies) by the desires of citizens. If citizens overwhelmingly want something, politicians are more likely to deliver (Gilens 2012; Lax and Phillips 2009).

Onward

In the rest of this book, we describe experiments conducted by ourselves and others that look at how people prevent disaster and climate change. As readers will see, with just the disaster game (and a few related games) there are countless ways to modify and extend the game to study a host of questions. Game research is only limited by the imagination of the researcher. If you can verbalize a strategic problem, you can almost certainly turn it into a game. Beyond helping readers understand the strategy of disaster, we hope this book inspires researchers to add games to their toolkit.

THREE

Dealing with Risk and Uncertainty

Can extinct animals live again? Michael Crichton's *Jurassic Park* converted us all into believers. The 1990 novel depicted a world where a mix of scientific know-how and the profit motive brought long-extinct creatures back to life. When *Jurassic Park* was published, de-extinction was science fiction. But decades later biologists have, if only barely and only briefly, achieved it.

The story centers on an unlikely star, the Pyrenean ibex, a large wild goat that roamed the Pyrenees. As of 1989, around a dozen remained; overhunting had decimated the population. About a decade later, the last known Pyrenean ibex, nicknamed Celia, was found dead under a fallen tree. But Celia would miraculously live on after death (Zimmer 2013). This was accomplished by a team of researchers led by José Folch, who used cells taken from Celia to create clonal nuclei and implanted these in goat eggs. Eventually, a single egg was brought to term by a hybrid Spanish ibex and goat. Celia's clone lived, though due to a severe lung malformation she died within minutes. De-extinction was real but fleeting.

The most obvious reason to de-extinct an animal is to restore an ecosystem. But can doing so also fight climate change? This seems unlikely on its face, but that's the attention-grabbing headline of Pleistocene Park (Andersen 2017). Located in Siberia, the park is run by the father and son team of Sergey and Nikita Zimov and serves as a testing ground for a unique type of geoengineering driven by raw animal power.

Until the last several thousand years, the Arctic was ringed with mammoth steppe, basically cold-weather grasslands. These grasslands were maintained by the actions of large herbivores—horses, reindeer, rhinos, and yes, mammoths. But due to some combination of human action and a changing environment, as the Pleistocene ended these mammals dwindled, and forests have increasingly replaced the grasslands. Unfortunately, compared to forests, the grasslands are better at maintaining permafrost. As the permafrost melts, it has the potential to release into the air massive amounts of carbon dioxide and methane—the two most detrimental greenhouse gases. Thus, recreating the mammoth steppe might curtail climate change. Besides keeping the permafrost frozen, such grasslands could directly lower temperatures because they have a high albedo and thus reflect light from the earth, slowing its warming.

But to recreate mammoth steppe, miles and miles of trees need to be knocked down and kept down, something mammoths do instinctively and effectively. If the Zimovs are right about the benefits of recreating the mammoth steppe, the most important science behind this geoengineering will not be about the climate directly, but about the science of genetics and reproduction. Will Pleistocene Park work? Although we don't know of any quantitative estimates, its success is certainly not guaranteed. And to the extent that the plan requires the de-extinction of mammoths (living species might be sufficient for the job), we would wager that many find the idea farfetched.

Pleistocene Park, as eccentric as it seems, illustrates a general problem with climate change: Aside from the basic premise that greenhouse gases are warming Earth, we know few things with certainty. How quickly is the world warming? What are the physical effects of the warming? How will it affect our lives and health? How effective will proposed solutions, like Pleistocene Park, be at mitigating climate change or allowing us to adapt to it? There are no easy answers.

Indeed, assessing and communicating uncertainty is a central concern of the Intergovernmental Panel on Climate Change (IPCC), one of the most important global authorities on climate change. It regularly releases reports that synthesize the state of the art research on climate change, reports that explicitly characterize uncertainty. IPCC reports describe the evidence along three dimensions: the amount of evidence, the quality of the evidence, and the extent to which the evidence points in the same direction (IPCC 2010).

For example, there is high confidence that precipitation from tropical cyclones will increase as global temperature rises. This confidence comes from a plethora of scientific studies, all converging on this claim. On the other hand, there is less confidence that precipitation will increase outside of these tropical cyclones as global temperature rises. Our question is:

What does this type of uncertainty mean for behavior? We will show, using the disaster game, how this sort of scientific uncertainty affects people's ability to prevent disaster.

Uncertain Impacts

How bad is climate change going to be? Based on popular media, you might assume unmitigated climate change is tantamount to the apocalypse. For example, one article in *New York Magazine* begins, "It is, I promise, worse than you think" (Wallace-Wells 2017). As we write this book, this article, titled "The Uninhabitable Earth," is the most read article in *New York Magazine* history. It describes heat waves killing people around the world, droughts that destroy food security, and even Siberian ice melting and releasing trapped bubonic plagues. Is this the future that awaits us if we fail to stop climate change?

The short answer is probably no. Scientists were quick to criticize the picture painted by the article. They argued that while there is some truth in its claims, the author cherry-picked only the worst-case scenarios (Vincent 2017). Instead, there is great uncertainty around many of the impacts of climate change, including its impacts on physical, biological, and social systems. In other words, there's a lot we do not know.

Does climate change make extreme weather worse? When describing climate change and the resulting storms, media reports often highlight the most extreme possibilities and leave out the uncertainty in the science of hurricanes and cyclones (Feldman et al. 2017). Hurricanes and cyclones form when moist air rises from warm ocean surfaces, creating an area of low pressure just over the water. Air from cooler, high pressure areas moves into the low-pressure zone, where it also warms and rises. This rising air creates clouds as it cools, forming a large, spinning system—a hurricane or cyclone.

The key ingredient for hurricane formation is warm ocean-surface temperatures, but it was initially less clear whether climate change has increased the number or severity of these storms. Some work says it has (Hansen et al. 2007). Other studies say that the number of severe storms has increased, but it is uncertain whether this is due to climate change (Webster et al. 2005). Finally, still other work says there has been no change in the frequency of hurricanes, but their destructiveness has increased (Emanuel 2005). In the context of this scientific uncertainty, in 2005 the director of the United States National Hurricane Center testified before Congress saying that climate change will not make hurricanes worse (Gray 2005).

More recent work has resolved some of the uncertainty; the 2014 IPCC impacts report synthesizes new information which suggests that climate change will indeed make hurricanes and tropical cyclones worse (IPCC 2014). However, even if scientists are now more confident that climate change increases hurricanes, there are still important unanswered questions. Where will we see the worst effects? How much can we attribute damages from different cyclones to climate change? How much can we expect hurricanes to increase? Ongoing work seeks to understand the future of hurricanes in the context of climate change (Bhatia et al. 2019; González-Alemán et al. 2019), and to determine how much each storm was exacerbated by climate change (Jézéquel et al. 2018, 2020), but there is still significant uncertainty in how climate change will disrupt these physical systems.

Of course, hurricanes are not the only extreme weather events that could be worsened by climate change. Climate change could, possibly, worsen tornadoes in the United States. Some studies suggest that climate change created more tornadoes (Agee et al. 2016; Elsner et al. 2015); other studies argue that although there are more tornadoes, this increase is not necessarily due to climate change (Tippett et al. 2016). There are similar arguments about whether or not we have seen or can expect more severe thunderstorms in the eastern United States (Diffenbaugh et al. 2013). These are, of course, just examples; our understanding of how climate change will affect many systems is bedeviled by uncertainty.

Using the Disaster Game to Study Mitigation

Because uncertainty is pervasive, we want to understand how people react to it. Does uncertainty about the impacts of climate change affect how willing people are to support mitigation? To answer this, we draw on research by ourselves and others that uses the *disaster game*. The disaster game is an economic game designed to capture key strategic features of mitigation. The original version was developed by Manfred Milinski and his colleagues (Milinski et al. 2008). We gave a brief overview of the game in Chapter 2 and now discuss it in detail.

Milinski and colleagues originally called the game the "collective risk social dilemma." Though the name is a bit intimidating for newcomers, it captures the essence of climate change. Climate change is what researchers call a *social dilemma*, which means there is a tension between the material interests of the group as a whole and of any member of the group.

In the real world, for instance, all nations would be better off if sufficient efforts are made to mitigate or adapt to climate change. This includes paying the costs of decarbonizing, providing adaptation aid to low-lying or island nations, and investing in noncarbon energy. However, except for perhaps the biggest nations (Smirnov 2019), no single country is absolutely required for these efforts to succeed. If, say, France decides to do nothing, the rest of the world could still solve the problem. What this means is that France could opt out of paying any costs—doing nothing to reduce emissions or rely more on renewable energy—and still reap the benefits provided by the rest of the world. It's not fatal if only France thinks like this. But if every country pursues a selfish course, then all nations will be worse off, compared to the case where all nations are willing to contribute. This is the crux of a social dilemma.

The game also captures the *collective risk* component of climate change. Changes to the climate affect everyone. America's emissions do not affect just America; China's do not affect just China. Although the problem is not spread in a fully even manner, it is certainly not a localized disaster like an earthquake or a flood.

Let's take an in-depth look at the original disaster game and remind ourselves of features common to most games. Milinski and colleagues recruited 180 students from the University of Cologne and the University of Bonn, in Germany, to come to the researchers' laboratories. As is typical, the players received a show-up fee just for participating, and this fee did not depend on anything else that happened in the experiment. They also earned additional money based on the choices they and others made in the game.

The students played in groups of six. However, they played through a computer network, working in semi-private cubicles with many six-person groups in the lab at once. Because of this, the students did not know which other students from their experimental session were in their group. By keeping players anonymous, the experimenter can reduce effects of any pre-existing relationships students might have with each other and reduce the ability of players to act later based on what happens during the game. The last thing experimenters want is for students, upset with others for failing to cooperate, to argue with each other in the hall! All these features—students in the lab, small groups of about four to eight players, and anonymity—are typical, though not universal, of games. All of these procedures, along with clear incentive structures, have been crafted by researchers over decades to ensure a carefully controlled environment (Ostrom 1998; Smith 1994). This allows researchers to make strong inferences about how changes in incentives and institutions affect people's decisions.

In this experiment, the students were each given €40. This real money is meant to represent tangible goods that could be lost to disaster and could be spent to prevent climate change. Each player's pot of money was theirs to control; no other player could influence the others' choices. When the game started, players were told that there was some risk that climate disaster would wipe out everyone's money. For instance, in one condition there was a 90% chance that disaster would occur. This means that for every ten groups who did not stop disaster, just one group would nonetheless keep their money. The fact that the fortunes of all players in the group are yoked together is why it's called a *collective* risk game.

So how could players prevent disaster? In Milinski and colleagues' game, the group was given a single *climate threshold*. The threshold was a monetary amount, in this case €120. If the players collectively contributed €120, then they averted disaster and kept any money they had left. Recall that there were six players who each started with €40, meaning each group started with €240. So, it cost half the group's resources to avert disaster. Players' decisions about contributing to the group unfolded over ten *rounds*. In each round, every player decided whether to contribute €0, €2, or €4 to the threshold. These contributions are meant to represent spending on mitigation, such as investment in emission-free energy like wind or solar power.

In each round, players made their contributions simultaneously, but after the round ended, they saw what everyone else contributed. They then moved onto the next round and again simultaneously decided how much to contribute. This continued until the tenth round ended. Based on their collective decisions, the group either averted climate change or still faced the original probability of disaster. The key strategic features of the game are illustrated in Figure 3.

Why a threshold, rather than every contribution chipping away at the total probability of disaster? One reason Milinski and colleagues set the game up this way is to represent climate *tipping points*. When a tipping point occurs, a dramatic change happens to the climate that could have sudden, swift, and disastrous effects for humanity (Lenton 2011; Lenton et al. 2008). (Tipping points have serious uncertainties associated with them; later in this chapter, we will look at how uncertainty in tipping points plays out in the disaster game. In the original version of the disaster game, and most subsequent variants, the climate tipping point is certain and known to all players.)

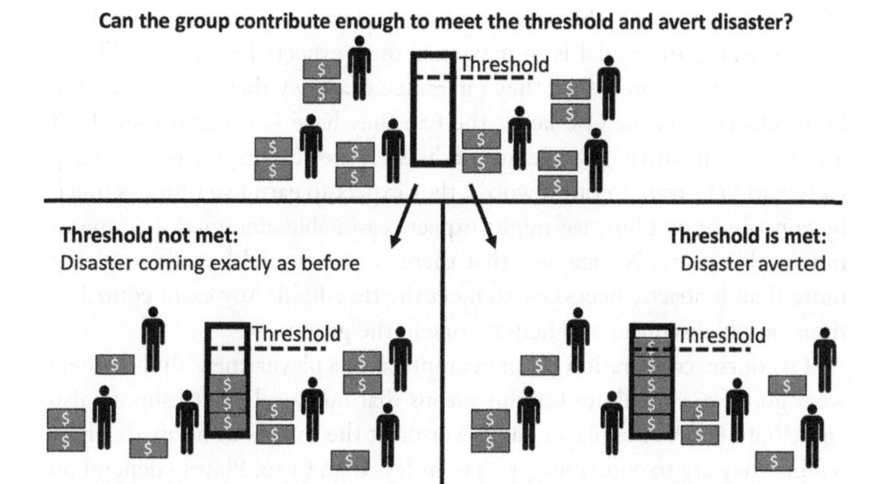

Fig. 3. The disaster game. Everyone starts with personal money (top panel). They each decide how much to contribute to a group threshold. If the combined contributions are less than the threshold, everyone loses their remaining money (bottom left panel). If the combined contributions are more than or equal to the threshold, everyone keeps their remaining money.

On average, if a group prevents disaster *and* everyone contributes equally, then each player must contribute €2 every round. As you might have noticed, this is where the *social dilemma* creeps in. There is nothing in the game that forces players to contribute equal amounts. Any particular player could contribute nothing in the hope that the rest of the group will solve the problem. This shirking player would end up with their original €40 when the game ends, whereas the rest of the players would end with only €16 on average.

On the other hand, the social dilemma aspect isn't as dire in the disaster game as in many games that capture social dilemmas. As we discussed in Chapter 2, the standard social dilemma game, called the public goods game, is set up in such a way that every player should defect no matter what the other players do. Things are different in the disaster game because success is defined by a stark, all-or-nothing threshold. Because of the threshold, players are best off if they coordinate around cooperation.

To see this, imagine that the game was played over only a single round and players each made a single decision about how much of their total pot to give (i.e., they can give any amount from €0 to €40). If one player knew that all the other players were each giving €20 (adding up to €100 of the needed €120), then the remaining player should also contribute €20 (assuming their goal is to maximize their expected earnings). This is because if they do give €20, they guarantee that they themselves and (as a by-product) everyone else keeps the €20 they have left. But if they don't give €20, then with a 90% chance of disaster, they can expect to only keep €4 (= €40 * (1 −90)). In other words, they expect to earn five times as much by contributing. Thus, we might expect reasonable amounts of cooperation in this game. Notice also that there is no material benefit to giving more than is strictly necessary to meet the threshold. Any extra contributions simply disappear and help no one in the group.

Of course, cooperation is not guaranteed. If a player knew all the others were going to contribute €0, this means that our focal player should also contribute €0. A lone player can never meet the threshold no matter how willing they are to contribute; €40 is far less than €120. Players depend on each other to work collectively to solve the problem.

As you can see, many different combinations of contributions by the players would add up to the threshold. What is best for one player to do depends on what the others are already doing. This feature makes the disaster game a coordination game. In this type of coordination game, success requires that the group coordinate its actions on a particular combination of choices that add up to the threshold. This is also what makes it different from the standard public goods game discussed in Chapter 2. In the standard public goods game, there is no coordination issue; everyone should defect no matter what everyone else does.

As we mentioned above, if the cost of meeting the threshold is shared equally, then each player should contribute half their original stake, that is, €20. This is often called the "fair-share contribution." Notice that in the basic game as Milinski and colleagues set it up, nothing within that game itself distinguishes the players—no one starts with additional money or faces an additional risk. Thus, if the players are generally cooperative and willing to contribute, there is no obvious reason why one player should contribute more or less than any other player. This makes fair-share contributions attractive as a *focal point*. Focal points help players solve coordination games because they are points among a sea of possible alternatives that stand out as obvious or intuitive in some way (Schelling 1960).

In Milinski and colleagues' setup, the players are indistinguishable: They are given the same amount of money to contribute toward climate change, and they all face the same risk of disaster. In later chapters, we study changes to the game that break this symmetry, such as giving players different wealth, different vulnerability to disaster, or different historical responsibility for climate change. For now, we hope you can see why in Milinski and colleagues' setup, where everything is equal, a series of fair-share contributions might be relatively attractive to players. We will often use the fair share as a benchmark.

Mitigation When Impacts Are Uncertain

Milinski and colleagues originally designed the disaster game to test how willing people would be to mitigate disaster depending on the *probability of unmitigated disaster*. Above we described the disaster game as involving a 90% risk of disaster if the threshold was not met. In fact, Milinski and colleagues manipulated this value, randomly presenting groups with a risk of 10%, 50%, or 90%. At a 90% risk, players who are in principle willing to contribute their fair share should in fact do so. As shown in Table 4, fair-share players can expect more money when all contribute than when all defect.

But things change for the other probabilities of disaster. Understanding this requires a brief detour into the idea of *expected earnings*. What does "expected" mean here? Consider this choice: Do you want $50 for sure or a lottery ticket that pays $100 half the time and $0 the other half? Although this lottery ticket will only pay $100 *or* $0, its expected value is what you would earn on average if you played the same ticket many, many times. That is, the lottery ticket's expected value is $50, the same as the sure thing. Despite this equivalence, you can probably see that most people are unlikely to be indifferent. Some people might prefer the sure thing, others the gamble.

So, what should players do at a 50% probability of disaster? Compared to 90%, they will probably be less motivated to meet the threshold. As the table shows, they can expect to earn the same regardless of whether everyone coordinates to meet the threshold or whether everyone defects.

TABLE 4. Earnings based on the probability of disaster

Probability of Disaster	Earnings with Fair-Share Contributions	Expected Earnings If All Players Defect
90%	€20	€4
50%	€20	€20
10%	€20	€36

Note: Based on Table 1 in Milinski and colleagues (Milinski et al. 2008). If each player gives their "fair share," they each give €20 of their original stake, exactly meeting the threshold, and leaving them each with a guaranteed €20. Expected earnings for defection are calculated as €40 * (1– the probability of disaster).

The only difference between these two possibilities is that if the group coordinates to meet the threshold, they keep their remaining money with certainty. But if they all defect, there is a 50% chance they each keep all €40 and a 50% chance they each end up with nothing. So, doing nothing is a gamble. When the expected earnings are the same, most people prefer a sure thing over a gamble. This means that players in the 50% version will probably want to coordinate, but not as strongly as in the 90% version—only in the 90% version is it definitely better to coordinate. (Players in the in 90% condition would have to be exceptionally fond of gambles to not want to meet the threshold, because the expected value of gambling on disaster is so low.)

Finally, when the risk of disaster is 10%, players who want to maximize their expected earnings should simply defect. Defection earns each player more on average than fair-share contributions. Even if players did not want to maximize their earnings, as a group they would have to be exceptionally averse to gambling to want to bother meeting the threshold.

Before looking at their data, take a moment to examine your own intuitions. What would your own first move be? If there were a 90% probability of disaster, what would you do on the very first round, before you see what any of the other players have contributed? Would you contribute €0, €2, or €4 to the threshold? Would you do anything differently if the risk were only 10%? If on future rounds, one or more of your groupmates seems to be contributing little or nothing, what would you do in response?

One of the beauties of games is that, even before any data is collected from players, they serve as a thought experiment about the strategy of the underlying problem. They force researchers to boil a real problem down to its essence. Notice how far we have travelled: from the real world of climate change and mitigation to a game that is simple enough to be understood and played by real people. If the process of simplification and abstraction has been done well (as we think it has been here), the game can reveal important facts about how people respond to this problem, in this case how they respond to uncertainty in the impacts of unmitigated climate change.

What do real players do? Let's start with groups. There were ten groups that faced a 90% risk. Of these, five groups successfully met the threshold and averted disaster. The remaining five groups came very close, contributing on average €113 of the needed €120 (just 6% short of the target).

Things change for the other two probabilities of disaster. For the ten groups facing a 50% risk of disaster, only one group met the threshold; for the ten facing a 10% risk, none were successful. And the 50% and 10%

groups contributed much less, on average €92 and €73 of the needed €120 (that is, only 78% and 61% of what was needed).

This is not bad news. Most groups responded reasonably. Only in the 90% risk condition is it clear that groups are better off stopping disaster, and these groups were successful or close to it. In the 50% condition, what players should do is murky. In the 10% condition, groups are probably better off not contributing; only if a group was full of people who really disliked gambles would it make sense to avert disaster here.

In fact, what's surprising is how much, rather than how little, players contributed in the 50% and 10% conditions. It's inefficient to spend a lot of money, but not enough to meet the threshold. Keep in mind, if the groups did not meet the threshold, disaster was coming exactly as before *and* the players permanently lost any money contributed to the threshold.

What do individual people do? In the 90% risk condition, about 60% of the time players gave their fair-share contribution of €2; the remaining contributions were roughly equally split between giving €0 and €4. So, in the condition where contributing clearly makes sense, players converge on the fair-share contribution most of the time. And doing so allows them to solve the dilemma (or at least get close).

Things change for the other probabilities of disaster. For the 50% risk groups, behavior does start out with roughly 60% fair-share contributions. However, it quickly drops off to only about 30%. Most of the players who didn't contribute their fair share instead contributed nothing. In the 10% risk condition, about half of all decisions were to contribute nothing. In these conditions, it appears that at least some players wanted to meet the threshold. But, after seeing other group members failing to contribute, they tended to back off from contributing themselves.

Recall that although it is less obvious in these conditions that the group should try to meet the threshold, it is not necessarily a bad strategy. It depends on how averse the players are to risk, a tendency that is likely to vary among players. So, the players who initially tried to contribute in the 50% and 10% conditions might have been ones who preferred a sure thing to a gamble.

In sum, Milinski and colleagues' disaster game presented a case where people must work together to stave off a catastrophic disaster. Some groups faced a disaster that was almost assured to occur absent costly prevention efforts (the 90% risk condition). Other groups faced a disaster unlikely to occur (the 50% and 10% conditions). This simple experimental manipulation captures an important part of the real-world uncertainties in the nature of the threat of climate change. How big a bite will climate change take if we do nothing? Is it an existential threat? A minor nuisance in the context of all the other problems facing humanity? Or, as most climate scientists think, is the problem somewhere—an uncertain somewhere—in the messy middle?

Room for Optimism When Impacts Are Uncertain

There are several takeaways from Milinski and colleagues' study. First, players generally responded in a broadly rational way to the risk of disaster. When disaster was nearly certain (90% risk), groups did a great job solving the problem. Indeed, if we think back to the real world, while actual tipping points are not perfectly smooth, they are not likely to be quite so all-or-none as the threshold of the disaster game. So, it's at least arguable that even the groups who just missed meeting the threshold were, for practical purposes, successful. At any rate, because the size of the real-world "threshold" to prevent tipping points can never be known with the certainty given in this game, close enough might be the most we can reasonably hope for.

A second takeaway from Milinski and colleagues' original study comes in the case when averting disaster makes sense (i.e., 90% risk). Here, players generally converged on fair-share contributions. This is important because, in the real world, coordinating large groups is complex, and the more obvious it is what each actor should contribute, the easier coordination is. At least when players are roughly equal with each other in terms of how much they can contribute, they find it obvious to contribute equal amounts. Later in the book we will discuss how well players coordinate when they are not equal to each other.

Third, and on a less optimistic note, players ended up wasting quite a lot of money in the cases where disaster was not actually serious (we will see this tendency again in Chapter 6). This suggests that people are overeager to avert disaster and may do so in ways that are economically harmful. In the real world, this might take the form of supporting policies or politicians who advocate for more drastic measures than are needed. To some extent, whether this is actually a problem depends on the true state of the world. For instance, proponents of the Green New Deal have likened solving climate change to fighting another world war, which would imply that extreme measures, including substantial government regulation of personal and economic life, are necessary to avert a worse moral horror. If elites accurately converge on this as the correct way to think about the problem, then voters will behave appropriately, since they are over-prepared to fight disaster anyway. On the other hand, Nobel laureate economist William Nordhaus argues that less invasive steps, including a carbon tax implemented appropriately and reasonably widely, will be sufficient (though of course not painless)—a far cry from fighting a war (Nordhaus 2013). If Nordhaus's view is broadly correct, then experts would need to do additional work to ensure that citizens to not go overboard. There are many issues to consider when determining what measures need to be taken. These include the trade-off between economic growth and mitigation efforts and how much we value future generations. These factors, which we address later in the book, are also subject to great moral and scientific uncertainties.

Uncertain Tipping Points

The idea of tipping points has a long pedigree. English encodes it in the cliché "the straw that broke the camel's back." Take a camel and add straw to his back, piece by piece. He won't even feel the first straw that's added. As a few hundred pieces are added, he begins to notice. As thousands go on, he feels the weight rise. But the impacts of the burden won't rise smoothly forever. At same point—the last straw—the load is too much and suddenly everything changes drastically and the camel's back . . . well, you know.

Tipping points are a major challenge of global warming; they are critical points at which the future state of the world is qualitatively and permanently altered (Lenton et al. 2008). Although there are potentially a wide variety of tipping points, scientists have often used the average temperature of the globe as a proxy for more specific problems. For example, according to the 2018 IPCC report, we need to reduce greenhouse gas emissions sufficiently to keep temperature rise below 1.5°C to avoid dangers from climate change (IPCC 2018).

While the IPCC has attempted to provide a clear benchmark for what needs to be done to prevent tipping points, there are (at least) three sources of uncertainty around this benchmark. The first uncertainty is how to define "dangerous," that is, which physical or social systems should be used as proxies. The second uncertainty is that we do not know exactly what rise in temperature will cause these dangerous impacts. Finally, there is uncertainty in how much we must reduce carbon emissions and take other mitigation measures to avoid this rise.

Which criteria should we use to determine when we have passed a tipping point? Timothy Lenton and colleagues compiled a list of fifteen different elements of the climate system that could be used to define a point of no return (Lenton et al. 2008), including the melting of the permafrost or the loss of Arctic summer sea ice. Which indicators we use will determine how much of a temperature rise we deem acceptable.

For example, one indicator that has passed a critical tipping point is the eradication of large-scale coral reef systems. Reef systems provide economic value to nearby communities and intrinsic value to the world because they are diverse biological systems. However, coral reefs can't survive large increases in ocean temperature. If the destruction of the coral reefs is our measure of dangerous climate change impacts, then we probably need to prevent temperatures from rising more than about 1°C (Baker 2001). We've already passed this degree of warming, so it is probably too late to avoid this tipping point (O'Neill and Oppenheimer 2002). (Of course, further uncertainty surrounds what effect reaching this tipping point might have on other ecological or social systems.)

Another possible criterion we could use is the destruction of the thermohalene circulation. The thermohalene circulation is a conveyer belt in the world's oceans. It pushes warm surface water up the European coast. The water then cools around Greenland and the Arctic, where it sinks and is pulled back down the North American coast. Climate change both warms the ocean and melts the northern sea ice responsible for propelling this conveyor belt. Extensive warming could cause this circulation to collapse. If the thermohalene circulation were to collapse, it would stop the flow of warm water up the European coast, which would cause dramatic cooling in northwestern Europe. Forecasts suggest that the circulation is unlikely to collapse unless there is a 3°C rise in global average temperature (Stocker and Schmittner 1997).

These are only two examples of tipping points; others involve human social systems (Oppenheimer and Alley 2004). Which tipping point we focus on suggests different rises in temperature that we need to avoid. As there is no clear answer about which indicators we should use, there is uncertainty about how low we need to keep the global temperature.

Unfortunately, even coordinating on a single indicator fails to resolve the uncertainty around how low we need to keep global mean temperature. For instance, one of the most accepted criteria of dangerous climate change is the melting of the West Antarctic Ice Sheet and the Greenland Ice Sheet. If these large masses of land ice melt, global sea levels might rise by five to seven meters (Revelle 1983). Estimates of the rise in temperature that would cause the melting of these ice sheets range from 1°C (Hansen et al. 2007) up to 4°C above preindustrial levels (Oppenheimer and Alley 2004; Oppenheimer and Alley 2005).

Finally, even if we coordinate on a single temperature rise to avoid, there is uncertainty over the reductions in carbon and other greenhouse gases that will accomplish this goal (Roe and Baker 2007). This is because elements of the climate system feed back and interact with each other in complex ways. The best current estimates—but they are only estimates—suggest that reducing CO_2 in the air to 350 parts per million (ppm) will keep the rise in temperature at approximately 1°C (Hansen et al. 2008). Concentrations of 450 ppm will lead to a 1.2–2.3°C rise, and concentrations of 550 ppm will cause between 1.5–2.9°C rise. The average concentration of CO_2 in 2018 was 407.4 ppm (Lindsey 2019).

Troubling Responses to Uncertainty in Tipping Points

As we discussed earlier, from the standpoint of cooperation, the threshold nature of the disaster game is helpful because it can transform the notoriously anticooperative setup of a typical public goods game—where everyone should free ride all the time—into a coordination game in which it makes sense for players to work together. Real players, as shown by Milinski and colleagues, can in fact be cooperative and successful in the disaster game.

But Milinski and colleagues made the threshold and the damage of disaster fixed and known to players. In the real world this is not the case. There is significant uncertainty in the exact location of tipping points. What do these uncertainties portend for the ability of people to cooperate? The most obvious concern is that when aspects of climate disaster are unknown, people will find it harder to coordinate and will therefore be less willing to cooperate. Is this the case?

Scott Barrett and Astrid Dannenberg (2012) created an experiment to find out. Illustrating the flexibility of games, their experiment departed in several ways from the original disaster game. In their version, players participated in groups of ten. Also, their players made a single contribution decision, rather than a series of decisions over multiple rounds. Moreover, unlike the original game, where failure to meet the threshold meant that there was some probability that *all* the money would be lost (say 10% or 90%), Barrett and Dannenberg's method worked differently. Here, if a group failed to meet the threshold, they definitely lost money, but not all of their money. Instead, a fixed amount was lost with certainty (though, as we'll explain below, the *size* of the amount lost was uncertain in some conditions).

Barrett and Dannenberg randomly assigned their groups to one of four conditions, with ten groups in each condition. In all groups, players started with two pots of money. First, they had €20, which they couldn't contribute to the threshold but could lose if their group failed to meet the threshold. Then, they had €11 divided into experimental tokens, so that each subject had a total of 20 tokens (the group had a total of 200 tokens). So, in total, each player started with €31 (and so the group had a total of €310).

Their *certainty condition* was similar to that in the original disaster game: Players had common knowledge of the exact amount it cost to meet the threshold, and the impact of failing to meet the threshold was certain and known. In this case, the threshold for a ten-person group was 150 tokens and if this threshold was not met each player lost €15 for sure from the money they could not contribute toward the threshold.

In the *impact uncertainty condition*, the size of the threshold was again 150 tokens. But if this threshold was not met, then the size of loss was uncertain. Rather than a fixed €15 loss, every player might lose anywhere between €10 and €20. The exact amount was randomly determined after the players had made their choices.

In the *threshold uncertainty condition*, the threshold itself was uncertain. Instead of a fixed and known 150 tokens, it could range anywhere from 100 tokens to 200 tokens. However, the impact of failing to meet the threshold was commonly known and was the same as in the certainty condition (€15).

In the *double uncertainty condition*, both the impact of disaster and the size of the threshold were uncertain.

Note that the expected value of these two features—the impact of disaster and the threshold—are equal to the values in the certainty condition: The average cost of disaster is always €15 per player and the average threshold is always 150 tokens. But whether the impact or the threshold is uncertain had very different effects on behavior.

Compared with the certainty condition, not knowing the exact impact of disaster did not make people less cooperative. In fact, under impact uncertainty players contributed a bit more. Each of the ten groups in the impact uncertainty condition contributed enough to avoid disaster, whereas only eight of the ten did so in the certainty condition.

Threshold uncertainty, however, greatly reduced cooperation. Compared to the certainty condition, players only contributed half as much. Even more ominously, nine of the ten groups contributed fewer than 100 tokens—ensuring that they would not prevent disaster even if they were lucky enough to face the minimum threshold amount. Players behaved similarly in the double uncertainty condition, contributing less as well. For instance, seven of the ten double uncertainty groups failed to contribute enough to meet even the minimum possible threshold of 100 tokens.

The results for *impact* uncertainty replicate the good news for climate cooperation found earlier: As long as the threshold is known, it is relatively easy for players to coordinate. On the other hand, *threshold* uncertainty drastically curtails cooperation. Why does one uncertainty have no effect while the other is debilitating? Barrett and Dannenberg argue that impact uncertainty does not change the underlying strategic problem of the disaster game. To the players, because they have a known target to reach, it remains a coordination game as before. And in this coordination game contribution still makes sense. Threshold uncertainty, on the other hand, reduces contributions because it removes the known threshold as a focal point and therefore removes a focal point around which the players can coordinate.

Is There Room for Optimism When Thresholds Are Uncertain?

So far, the experimental evidence suggests that threshold uncertainty is bad for cooperation. But in Barrett and Dannenberg's initial study, the uncertainty about the true value of the threshold was quite large, ranging from 100 tokens to 200 tokens. Maybe people can tolerate uncertainty, but only up to a point. In a follow-up study, Barrett and Dannenberg (2014) again introduced threshold uncertainty. As in their original experiment, the follow-up included a certainty condition and a condition where the threshold ranged from 100 tokens to 200 tokens.

The important change is that they added three more conditions, each with a progressively smaller range for the threshold: 135 to 165 tokens, 140 to 160 tokens, and 145 to 155 tokens. Across the five conditions, then, all the ranges were centered on 150 tokens, but the size of the range was either 0 tokens (i.e., certainty), 10 tokens, 20 tokens, 30 tokens, or 100 tokens. Barrett and Dannenberg created a game theory model of this setup and found that it predicts that players should prevent disaster in the certainty condition and when the range is only 10 tokens. When the range is bigger, players should fail to meet the threshold.

What do real players do? Can they sustain cooperation in the face of some, though not overwhelming, uncertainty? It seems that they can. As in their previous study, each condition consisted of ten groups of ten players. First, and not surprisingly, in the certainty condition eight of ten groups met the threshold. More importantly, when the range was just 10 tokens, players continued to be fairly successful. In this condition, eight groups prevented disaster for a full 60% of the range of possible thresholds. This means that each of these groups contributed 151 tokens or more. Recall that the range is centered on 150 tokens, so these players were pretty close to it. (Sixty percent of the range from 145 to 155 tokens is 145–151 tokens.) This success occurred even though players were ignorant of the exact threshold. Reinforcing the similarity between behavior in the certainty and 10-token range conditions, players' contributions were not statistically different between these conditions.

In the other three conditions, which had greater uncertainty, twenty-nine of thirty groups failed to contribute enough to meet even the minimum possible threshold in their condition. And the lone outlier group avoided disaster for only 30% of the range of possible thresholds. Still, even in these conditions, contributions were greater than the zero level of contributions predicted by Barrett and Dannenberg's game theory.

Altogether these results are reassuring. While players can solve the problem when the solution is absolutely certain, in the real world we are never likely to eliminate all uncertainty. However, uncertainty is not invariably fatal. When players face a hazy though still reasonably defined target, they are about as successful at meeting it.

This finding is also useful for research using the disaster game. Knowing that some uncertainty does not change the game much from complete certainty makes running experiments easier. If uncertainty always had to be included, the game would quickly become too complicated when researchers layer other elements on top, such as adding differences between players in wealth or in risk. Because players play the same when facing certainty and when facing (some levels of) uncertainty, researchers can use the certain version of the game without sacrificing generality.

Finally, this finding informs the real-world program of communicating the scientific uncertainty in climate tipping points. On the one hand, we need to accurately describe uncertainty to citizens so that they can make informed decisions. On the other hand, if too many citizens perceive too much uncertainty, then the potential for coordination unravels. This makes resolving climate uncertainties a priority.

Which Mitigation Strategies Will Work?

Will Pleistocene Park make a big difference in fighting global warming? If you think so, how much would you bet on it? Pleistocene Park is certainly a high-risk, high-reward idea. If you're like us, you might think the idea is ingenious but not likely to work out in the end. Of course, there's nothing wrong with this. Solving a large and difficult problem is going to involve a lot of blind alleys and out-there ideas. Some will pan out, some won't. The trouble is that we—everyday people, policymakers, scientists—do not have infinite time or resources and need to decide carefully where to place our bets.

With its benchmark of keeping global warming below a 1.5°C rise, the IPCC emphasizes a shift toward energy efficiency, changes in infrastructure and mass transportation, and more reliance on renewable energy sources such as wind and solar power (IPCC 2018). All of these strategies will gradually reduce the amount of carbon dioxide released into the atmosphere. However, because the world so far has resisted serious mitigation efforts, it is unlikely that these certain, incremental technologies will be enough to avoid dangerous climate tipping points.

Another path described by the IPCC is the creation of technology that removes carbon dioxide from the air, called "negative emissions technologies." Some of this "technology" is as simple as afforestation—planting more trees. Others are more complex, such as technology that captures and stores carbon. (These technologies are sometimes called "carbon scrubbing.") These strategies all demand difficult tradeoffs; for example, afforestation competes with agriculture for land. But, more importantly for our purposes, they introduce uncertainty about how effective they will be, if they are even effective at all.

Many of these technologies are uncertain in that they simply do not yet exist on a scale at that could successfully mitigate climate change. Money and time invested in these new, uncertain projects would be wasted if they fail. The IPCC mentions strategies like directly capturing carbon from the air, alkalizing the ocean, and sequestering carbon in the soil; for none of these do we know the exact probability of success. A challenge for policy making is to know whether and when to invest in high- or low-risk technologies. One part of this is learning what kind of investment real people are likely to choose.

Choosing between Mitigation Strategies in the Disaster Game

We designed a series of games to test this (Andrews et al. 2018). One thing we haven't yet mentioned about the disaster game is that in many uses the game is *framed* (the original game by Milinski and colleagues was a framed game). In a framed game, the players are told to interpret the game in terms of a real-world problem. In our game, this means that we explicitly told players that the game was designed to capture how people might work together as a group to solve dangerous climate change. We also told them that each person in a group would have to decide whether to invest in high-risk, high-reward or low-risk, low-reward mitigation technology, explicitly contrasting technology such as solar or wind power with carbon scrubbing.

There are several virtues to this. First, a real-world example helps players understand the game. Games are not like public opinion surveys. A typical survey question is brief, such as "How serious a threat do you think global warming is to you and your family?" (O'Connor et al. 1999). Games typically require several pages of instructions. Usually, basic math skills are required of players—at the very least some simple arithmetic and a rough understanding of probabilities. Given this complexity, anything that can help the players understand the game is useful.

Second, the same abstract game (in a researcher's eyes) might be thought of very differently by real players depending on what real situation the players view or are told the game applies to (DeScioli and Krishna 2013). For instance, a classic economic game, called the dictator game, involves two players. One player is given, say, $10 and gets to unilaterally dictate how the money is split between the pair. If the game is played between perfect strangers, players might construe it one way. But if it's played between two family members, the game will probably be perceived as different in its underlying nature. It's not just that players will be quantitatively more generous (though they probably will); it's that they will apply an entirely different mental frame to the encounter (Fiske 1992). Think of your own life. Is a stranger you fleetingly interact with in the supermarket merely different in degree from your parents, siblings, spouse, or children? Most of us would answer no: Our intimate relationships are of an altogether different kind. With this in mind, by framing the game in terms of the real-world issue we are interested in replicating, we can make sure that any mental frames the players bring to the game are the same as what they would bring to thinking about the real-world issue.

In one of our studies on investing in climate technology, we brought 296 Stony Brook University undergraduates into our lab to play a modified version of the disaster game. They played the game in groups of four. When the game began there was a 90% chance disaster would destroy everyone's money. Each player made a single decision about how to invest their personal money toward the climate threshold, and then the game ended. Players made their decisions simultaneously, so they did not know what the others decided.

Every player started with 100 experimental tokens. (Each token was worth 20¢, for a total of $20). However, the 100 tokens were divided between two pots. One pot was worth 80 tokens and was designed to represent infrastructure—things like roads, bridges, and buildings that could be damaged if disaster happened but cannot be meaningfully used to prevent disaster. The remaining pot of 20 tokens could be used to invest toward the climate threshold. This pot was how we represented investment in high- or low-risk technology. Players could either keep all of this money or they could contribute all of the pot toward one of two technologies (i.e., mixed investment portfolios were not allowed, to keep things simple).

If a player did decide to contribute toward the climate threshold, one option was to contribute directly, applying 20 tokens toward the threshold with certainty. This represents investing in technology like solar or wind power. The other option was to make a risky contribution, essentially gambling the money. If the player chose a risky contribution, then there was a 50% chance that 40 tokens would go to the threshold and a 50% chance that 0 tokens would. This represents investing in technology like carbon scrubbing. Thus, this choice is high-risk, high-reward, compared to the certain contribution. At the same time, both contributions have an expected value of 20 tokens. If multiple players in a group invest in risky technology, their gambles are independent—if one player's gamble pays off, that says nothing about whether another player's will.

Our primary question was whether players would make sensible decisions in their investment choices based on the difficulty of the problem they faced. To do this, we varied the size of the climate threshold. For instance, in one condition, groups faced a threshold of 80 tokens to prevent disaster. There are multiple ways this threshold could be reached. For instance, all four players could make a risky investment and hope that at least two gambles pay off (recall that a successful gamble is worth 40 tokens). About 69% of the time, at least two players would succeed, and the disaster would be averted. Or two players could make certain contributions (recall that each certain contribution is worth 20 tokens) and the other two could both gamble. So long as at least one gamble pays off (a 75% chance), the disaster is also averted.

But why risk it? At a threshold of 80 tokens, the four players can guarantee that they avoid disaster if all players make a certain contribution. In fact, in a game-theory analysis, when the threshold is 80 tokens, we found that the best thing for a group to do is for each player to contribute their 20 tokens as a certain contribution. This guarantees that they each keep their remaining 80 tokens, and this is better for each player than everyone defecting. If everyone defects, each player can expect to earn only 10

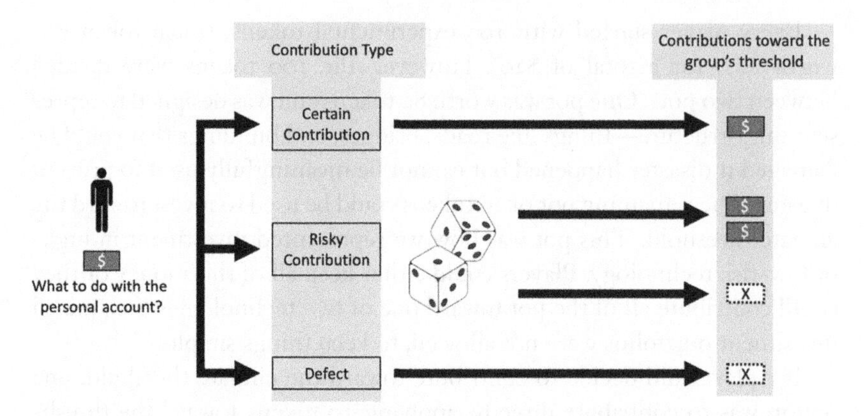

Fig. 4. The decision each player faced. A player starts with a personal pot of money. If they make a "certain contribution," that money is given directly to the group's threshold. If they make a "risky contribution," there is a 50% chance the money will be doubled before going to the group's threshold and a 50% chance the money will disappear and nothing will go to the group's threshold. If they "defect," then they keep their personal money, and nothing goes to the group threshold.

tokens (their original 100 tokens multiplied by the 10% chance that they avoid disaster even if they do not meet the threshold). Remember also that the disaster game is a coordination game. Thus, so long as all the other players are contributing, the final player is better off contributing rather than defecting. Altogether, we would expect that when the threshold is 80 tokens many players will choose to make a certain contribution.

Things change if we consider other thresholds. In another condition, groups faced a threshold of 120 tokens. Obviously, this threshold cannot be met by just certain contributions alone (120 tokens > 4 * 20 = 80 tokens). Necessarily, at least some players must invest in risky technology *and* have those gambles pay off. Their best shot is for all four players to make a gamble and hope at least three pay off. There is about a 31% chance that at least three do. This means that if all four players make the risky contribution, they can expect to keep 31 tokens, compared to only 10 tokens if no one contributes. So, when the challenge is greater, players cannot necessarily expect their groups to be successful all the time—even when making the best possible decisions—but contributing is still better than not contributing. Thus, we would expect that when the threshold is 120 tokens, many players will choose to make risky contributions.

Altogether, we studied five threshold sizes. Two of them could be met with certain contributions: 60 and 80 tokens. For the 60-token case, just three certain contributions will guarantee that disaster is prevented; the final player can defect as they are strategically superfluous (lucky them). As described above, the best thing at 80 tokens is for all players to make a certain contribution.

The other three threshold required at least some successful gambles: 100, 120, and 140 tokens. It turns out that the number of required risk-takers does not smoothly rise with these thresholds. At 100 tokens, there are two equally efficient sets of decisions. The first has one player make a certain contribution and the remaining three gamble (hoping at least two pay off). The second reverses this with one player making a risky contribution (hoping it pays off) and the remaining three making certain contributions. So, on average between these two sets, two out of four players should make risky decisions. At 120 tokens, as described above, the most efficient set is for all four players to make a risky contribution. Finally, at 140 tokens the best thing to do is for one player to make a certain contribution and for the remaining three to make risky contributions (and hope all three pay off). We were not sure how sensitive players would be to these subtleties, but we kept our eyes open for them. Our more general expectation was that players would tend to make more risky contributions at the three highest thresholds, where at least some gambling was necessary—compared to the two lowest thresholds, where they could be met by certain contributions alone.

People Know When to Take a Risk

How did our players respond? To begin, we found that players in this game were very cooperative: Only about 10% did not contribute at all, and this defection rate did not depend on what threshold the group faced. In Figure 5, we graph the proportion of players making risky contributions. Players made relatively few risky contributions at the two low thresholds, but they made relatively more at the three highest thresholds. (Because the same number of players, approximately 10%, defected at all thresholds, the proportion of players making *certain* contributions is simply what is left over after accounting for the proportion making risky contributions, minus 10%.) Altogether, players in this game were very cooperative and generally responded to the size of the threshold in a sensible way.

We were concerned, however, that a feature of our game may have pushed players to be cooperative in a way that might not be realistic. The infrastructure pot of money was four times larger than the liquid pot.

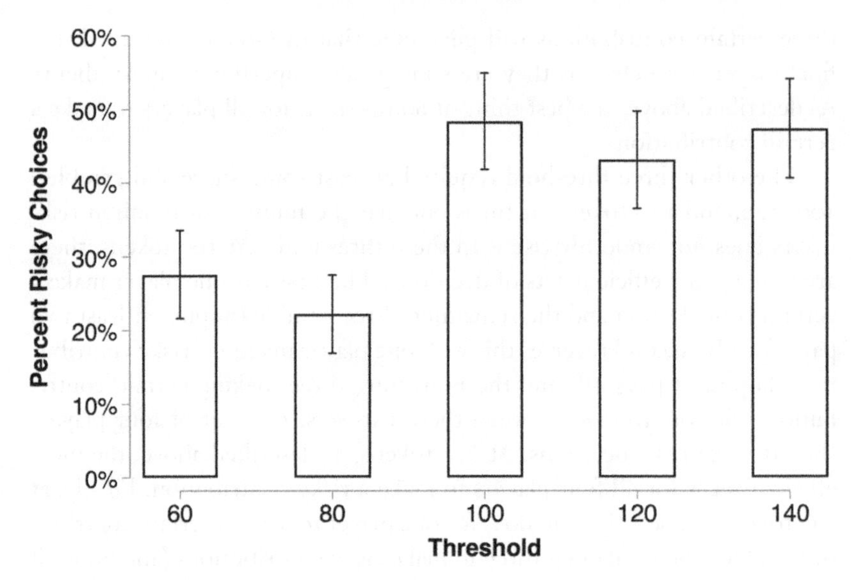

Fig. 5. The percentage of players making risky contributions in the lab with high stakes. Error bars are standard errors of the mean.

Maybe players saw themselves as having too much to lose and couldn't help but contribute.

To see how much this affected our players, we ran a nearly identical version of the experiment with 297 new undergraduate players. The only difference here was that the infrastructure pot was worth only 40 tokens in this low-stakes version (that is, cut by 50% from the original 80 in the high-stakes version); the liquid pot was still 20 tokens. Despite this change, players behaved nearly the same. Again, players were very cooperative, with only about 8% not contributing. As Figure 6 shows, players again made relatively few risky contributions at the two lowest thresholds and relatively more at the highest thresholds. Here, too, players were cooperative and quite sensible in their decisions.

To check how robust these results were, we replicated both these studies on new samples. Instead of using more students, however, we turned to an internet sample. This sample comes from Amazon's Mechanical Turk service, usually shortened to MTurk. MTurk connects people who need many brief tasks done, like categorizing a photo as containing a human face or not, with people willing to do these tasks for small sums of money. Many MTurk workers are also happy to complete, for pay, brief social science studies. One benefit of having MTurk workers complete the study is that they are far more diverse than the student samples (although they are

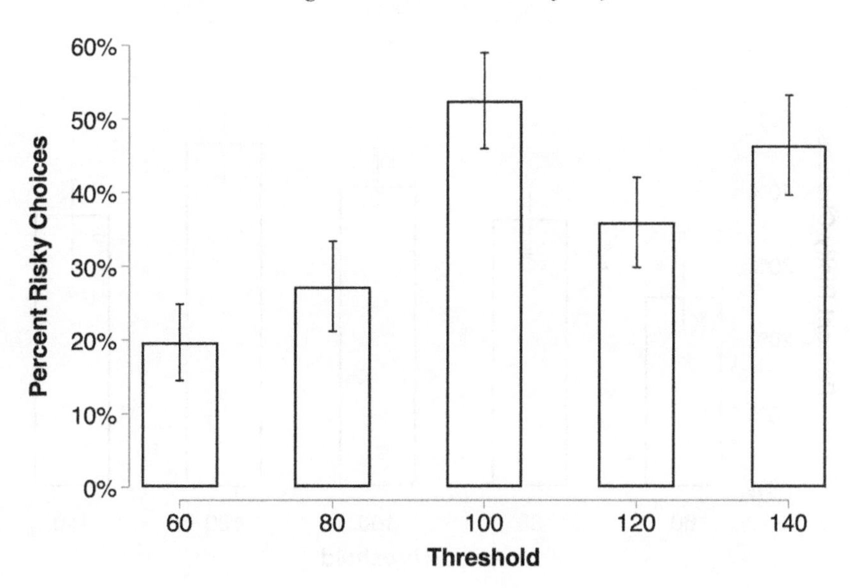

Fig. 6. The percentage of players making risky contributions in the lab with low stakes. Error bars are standard errors of the mean.

certainly not a representative sample; see Berinsky et al. 2012; Buhrmester et al. 2011). On the other hand, the downside is that they are not physically in our lab, and we therefore have less control over their experimental experience.

One important difference between laboratory and MTurk economic games is the size of the stakes. MTurk stakes are much smaller, often around $1. Perhaps surprisingly, past research shows this does not substantially affect behavior (Amir et al. 2012). In our MTurk experiments, players again played with tokens, but each token was worth 1¢. Thus, the total stake was $1 in the high-stakes version (which had 501 players) and $0.60 in the low-stakes version (which had 499 players).

Despite the big changes in sample and setting, gameplay was similar in these studies. Again, only about 10% of players defected. As shown in Figures 7 and 8, the proportion of players making the risky contribution was generally higher at the three highest thresholds. The obvious difference between the MTurk and lab samples is that MTurk workers were far more sensitive to the exact value of the threshold. Their behavior mirrored the complex pattern predicted by a close analysis of the game theory: risk-taking rising up to a point but then falling a bit at the very highest threshold. We were surprised by this because we generally have no idea where

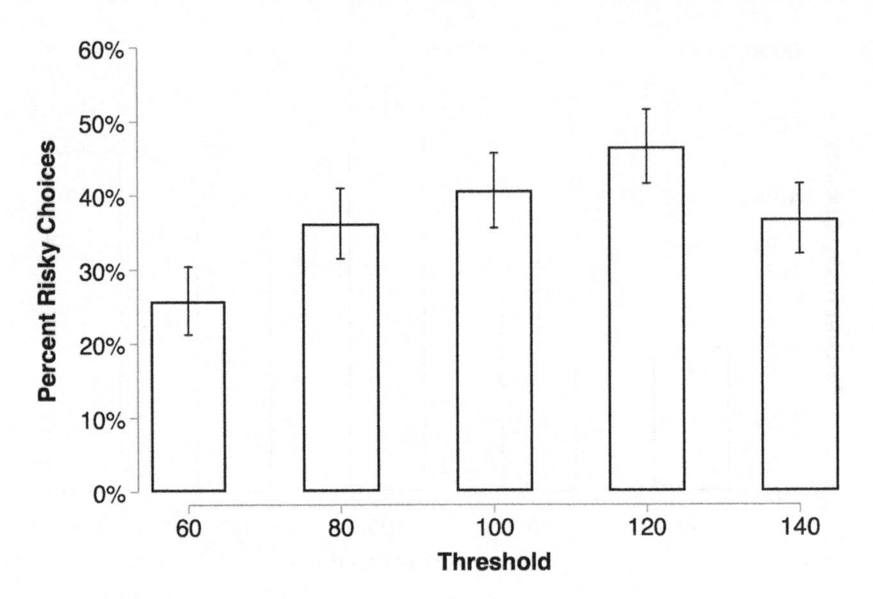

Fig. 7. The percentage of players making risky contributions online with high stakes. Error bars are standard errors of the mean.

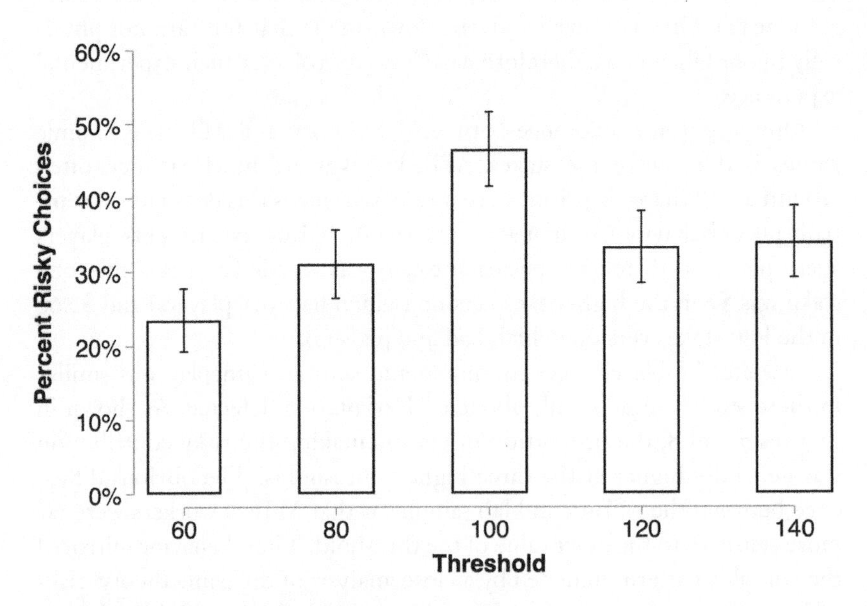

Fig. 8. The percentage of players making risky contributions online with low stakes. Error bars are standard errors of the mean.

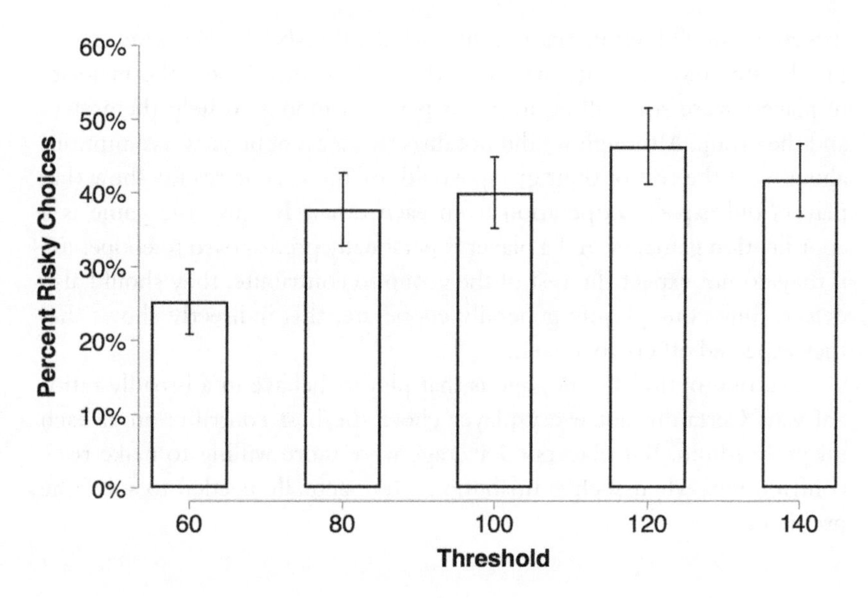

Fig. 9. The percentage of players making risky contributions online with high uncertainty in the risky contribution. Error bars are standard errors of the mean.

MTurk workers complete studies—they might be at home, on a bus, in a club, at the park, or anywhere. And the stakes are much smaller. For both these reasons, one might think MTurk workers would pay less attention. But in fact, if anything, they were better game theorists than our undergraduates. One reason we suspect this might be the case is that the typical MTurk player in our game is more experienced in taking studies than the typical undergraduate lab participant (Hauser and Schwarz 2016).

We ran one final study on MTurk. The previous studies had a stark gamble: Either your wager on technology is doubled or it disappears entirely. Real-world technology will surely have a higher variance in its effectiveness. To capture this, in this final study there was still a 50% chance that an investment would fail outright (most new technology is unlikely to get out of the gate). But if it doesn't fail outright, there is an equal chance that it ends up contributing 80, 60, 40, 20, or 0 tokens to the threshold. Despite the higher variability, the gamble still has an overall expected return of 20 tokens. As shown in Figure 9, players in this version continued to make more risky contributions as the threshold increased.

What does our players' behavior tell us? We think it reveals two optimistic messages. First, our players were quite willing to cooperate; recall that only around 10% defected in these studies. Also important is that the defection rate did not increase along with the threshold. Even when averting disaster was difficult, that is, at the highest thresholds, the majority of players were still willing to invest personal money to help themselves and the group. Although we did not directly measure players' assumptions about what the rest of their group would do, these same results show that players did expect cooperation from each other. Because the game is a coordination game, even if a player is personally predisposed to cooperate, if they do not expect the rest of the group to contribute, they should also defect. Since our players generally cooperate, this indirectly shows that they expected others to as well.

A second optimistic message is that players behave in a broadly rational way. Certainly not every player chose the best contribution at each given threshold. But players on average were more willing to make risky contributions when such contributions were actually needed to solve the problem.

Leveraging Evolved Psychologies for Climate Change Mitigation

Something that we have not mentioned yet is that we designed this game not just to understand a real-world problem, but with a specific psychological heuristic in mind. Research on this ability—what's called "risk-sensitive decision-making"—began several decades ago in evolutionary biology, studying nonhuman animals in contexts far removed from disaster prevention. The original venue was testing whether foraging animals would be sensitive not just to the *expected value* of foraging at some location but also to the *variability* associated with each location.

An early series of experiments studied a bird called the yellow-eyed junco (Caraco et al. 1980). In these experiments the juncos were given a choice between two feeding stations. For example, in a trial one station might give the bird two seeds with certainty and the other would give zero seeds half the time and four seeds half the time. That is, the expected number of seeds was constant, but the riskiness varied.

The researchers wanted to know how the birds' caloric needs (analogous to our climate threshold sizes) influenced their willingness to take risks. In one condition the birds were not given food for one hour prior to testing, which made them hungry but does not unduly tax their energy needs. In 76% of trials, these birds preferred to take the safe option. And no bird was actively risk-seeking here: In the remaining trials, the birds were merely indifferent. These sated birds felt no need to gamble.

In the other condition, however, the birds were not fed for four hours. This is quite taxing and means the birds were starving—unlike us, these birds have to consume a lot of calories every single day to simply survive the night. Here, behavior flips, and the birds preferred the *risky* foraging patch in 62% of trials, and in the remainder were simply indifferent. For a starving bird, the small amount of food, though guaranteed, is less likely to do the trick. Only by taking a risk do they have a reasonable shot at surviving. (At least, in a natural environment; the researchers of course would not actually have let the birds starve.)

As this study vividly illustrates, survival often depends on the ability to make good decisions integrating need and risk, so it is no surprise that this skill is found in animals as diverse as insects, fish, birds, and mammals (Bateson 2002). This ability is also found in people (Mishra 2014; Rode et al. 1999; Wang 1996). Thus, the problem of deciding what kind of technology to invest in seems tailor-made for our psychology for managing risk.

It might be that many decision problems, once stripped of extraneous details, actually map onto the kinds of problems that humans have heuristics for solving. In fact, we and many others have argued that humans also have heuristics for successfully working together collectively and for creating public goods and other collective benefits (Delton et al. 2013; Delton and Sell 2014). Moreover, these abilities, originally tailored for small-scale social life, might also guide our thinking in mass politics.

Thus, scientists and policymakers might help average citizens by making sure that information about climate problems is presented in ways that intuitively click for people. For instance, the psychologist Gerd Gigerenzer has argued that the human mind is not equipped to easily handle probabilities when they are expressed as percentages (Gigerenzer 1998; Cosmides and Tooby 1996). The mind is, however, good at handling what are called "natural frequencies." For instance, when a commercial says that three out of four dentists recommend a new toothpaste, the advertisers are using natural frequencies; had they used percentages, they would have said 75% of dentists. (Incidentally, this would explain why many psychology experiments seem to show that people are bad at probabilistic reasoning: It's not that people are actually bad at probability, but that the information is not being conveyed in a useful way, that is, in a way that works with our evolved psychology.)

Simply presenting information about disaster in a way that the mind is better at using is likely to lead to better decisions by citizens. Our experiments on risk do this by combining two types of problems that humans are routinely good at solving: group cooperation and risky decisions.

Responding to Scientific Uncertainty

How difficult and how disastrous will climate change be? While the media often cherry-pick worst-case scenarios, there is a lot of uncertainty in the science. But this raises the question: How will people respond to these uncertainties?

Based on research using the disaster game, we think there is in fact plenty of room for optimism. Players respond sensibly to the risk posed by unchecked climate change (Milinski et al. 2008). Although players have difficulty cooperating when there is considerable uncertainty, a little uncertainty does not preclude success (Barrett and Dannenberg 2012, 2014). And players understand when low-risk investments in mitigation technology are best and when high-risk investments are (Andrews, Delton, and Kline 2018).

One feature that united all the games in this chapter is that players made decisions about preventing disaster for themselves. In the real world, however, we often make decisions about preventing disaster for others. We turn to this problem next.

FOUR

Deciding for Others

The island nation of Kiribati (KEE-ree-bas) spreads its more than thirty atolls over the middle of the Pacific Ocean, ranging across nearly 1.5 million square miles of water. Its land mass, however, encompasses a tiny 300 square miles and holds a population of just over 110,000 (Factbook 2019). The people of Kiribati, few in number and with a per capita GDP of $2,300, have contributed nearly nothing to global emissions—their population is essentially a rounding error next to the global population, and they consume far less than people in developed countries.

Yet scientists project that Kiribati will be submerged by the year 2100 (Ives 2016). The habitable areas of the islands rise only 2 to 3 meters above sea level. Due to climate change, seas throughout the world could rise as much as 2 meters. This, combined with local sea level changes and weather events around Kiribati, has the potential to send the island nation underwater (Donner and Webber 2014). Although an extreme case, Kiribati provides two examples of what happens when climate decisions are made for some people by others. First, of course, the citizens of Kiribati emitted almost nothing, yet they will face enormous consequences from others' emissions. Second, their attempts to adapt illustrate how hard it is for distant, unaffected others to provide useful help.

Let's dig deeper into the latter. Given the short timescale, the rest of the world may be impotent to avert the sea level rises that threaten Kiribati. So, the nation must choose between some type of local adaptation or abandoning their islands entirely. Indeed, the country has already purchased land in Fiji in case they must move.

Adaptation is difficult for many reasons. One difficulty that bedevils adaptation is uncertainty. The exact rise of sea level, globally and locally, is not known with precision. In 2005, Kiribati's own government projected only 0.1 to 0.8 meters local rise by 2100; other groups later estimated that the rise could be as high as 2 meters. The islanders cannot easily plan how to adapt, or chose between adapting and leaving the islands, with such a large range of possibilities.

Adaptation is also difficult because the people of Kiribati could use multiple strategies to prevent damage. One possibility is seawalls, which come with tradeoffs. For instance, the type of seawall that can be easily constructed by the local population will erode the island. An alternative and less complicated strategy is to plant mangrove trees. There are not exactly downsides to mangroves, because they would prevent erosion and may even cause land growth. But using this solution alone might not solve the problem, so only planting mangroves could take time away from other adaptation strategies.

Yet another difficulty, and our focus in this chapter, is that the citizens of Kiribati are not making this decision on their own. Because Kiribati is poor and not responsible for its plight, outside governments and other organizations have been providing money and advice. This includes the World Bank, which has used Kiribati as a "demonstration project." In conjunction with the Kiribati government, the World Bank formed the Kiribati Adaptation Project, and over its first several phases invested nearly $20 million on adaptation.

The seawalls that resulted from this project would later be called "embarrassing," "built wrong," and "poorly done," adding to a "graveyard of short-lived infrastructure investments" (Donner and Webber 2014). This falls far short of what the citizens of Kiribati wanted: projects that will protect their grandchildren and great-grandchildren. How did this happen? First, outside consultants often operate at a distance and report to even more distant bosses and funders. Even if outsiders know what the citizens of Kiribati want, they tend to have different goals. They need progress now so they can show something for the money and time invested—hence the slapdash seawalls. Even projects that look good right away are a waste if locals cannot maintain them due to lack of knowledge or funds—hence the infrastructure graveyard.

Rational Abstention and Rational Ignorance in Politics

A general problem lurks here, one that complicates not just climate decisions but many political decisions. In fact, before getting to deciding for *others*, it's helpful to see the problems that arise when we make decisions that affect *ourselves*. Consider a national election. Voting costs you something; it takes time and energy to go to the polls and cast your vote. At the same time, your vote is just one of many and there's virtually no chance it will decide the outcome of the election. The only way that your vote matters is if there would have been a tie without you. With no tie, you might have well stayed home. And in a large electorate, the probability that you will need to break a tie is basically nil. This means that if you only care about yourself, then you should not waste your time or energy voting.

This disturbing logic was pointed out several decades ago by the political economist Anthony Downs (1957; see also Buchanan and Tullock 1962). Downs noticed that in large elections, the chance that you cast the decisive vote is vanishingly small, so it does not pay to vote *even if you personally have a lot at stake*. Suppose you expect to be $5,000 richer if Candidate Jones beats her opponent. Shouldn't you vote for Jones and get that money? Not really: If your vote is unlikely to affect whether Jones takes office, there's no point in bothering. It's great for you if Jones wins, but you cannot realistically help. If you decide to stay home and not vote because of this, it's called *rational abstention*. (Of course, if *everyone* else stays home, and you alone vote, then you decide everything. So, Downs's theory does not predict that literally no one should vote, just that only a few should.)

The problem gets even worse: Downs recognized that there's also a problem of learning. If a rationally self-interested person will not bother to vote, why should they learn about the issues or candidates? After all, if the voter won't benefit from their voting, they certainly won't benefit from taking the time to learn who or what to vote for. Thus, Downs's theory predicts not just rational abstention but also *rational ignorance* of politics. Why learn about boring topics like water treatment, road repair, city budgeting, or climate change if you will never vote in an election about them? (This is true if we think of people learning about politics strictly as a means to an end. But some people enjoy politics the way others enjoy sports and so treat politics as entertainment. For the latter camp, it would make sense to learn about politics, the same way it makes sense for sports fans to learn about their favorite players; Prior 2005.)

Consistent with the prediction of rational ignorance, studies often find that voters don't know much about politics (Delli Carpini and Keeter 1993). A recent review found that 44% of Americans in 2013 did not even know that the Affordable Care Act (aka Obamacare) was still the law of the land. In 2014, only half the country knew that "Common Core" had something to do with education. Also in 2014, over 80% of the country did not know that the federal government spends less than 5% of its budget on foreign aid; about half of this ignorant set thought the US spends ten times or more than it actually does (Somin 2016). Voters systematically misunderstand how government policies affect their welfare; for example, they underestimate the benefits of foreign trade on economic growth (Caplan 2011).

What are the implications for climate decisions? Choices about how to prevent disasters are political decisions, and there does not seem to be a rational reason for people to expend energy to make these choices, even if they are personally affected by the disaster. With climate decisions, of course, it's going to be one set of people making decisions on behalf of a different set of people. Current generations are going to make decisions for future generations (Nolt 2019). Rich, developed countries are going to make decisions that affect vulnerable poor and developing countries (Rozenberg and Hallegatte 2018)—just think back to the Kiribati example. If there is no value in becoming informed enough to make ourselves better off through politics, what hope is there that people will learn about or think through a problem facing distant others, including those not yet born?

Social Preferences and the Possibility of Making Good Decisions for Others

This is dispiriting, but on the other hand it doesn't fully capture reality: Lots of people do vote, and voters do seem to discipline their governments. Why might this be the case? First, people don't care only about the benefits they receive if the policy or candidate they prefer wins the election. Some voters get benefits just by the act of voting. For example, they might feel good about fulfilling their civic duty to vote (Riker and Ordeshook 1968). They might also vote to enjoy the camaraderie of voting with friends (Rolfe 2012) or to avoid the shame of admitting they didn't cast a ballot (Dellavigna et al. 2017).

A second problem with the pessimistic analysis above is that it assumes people are only interested in their own benefits. But people often care about what happens to others, too. Economists and psychologists have produced reams of data showing that people have interests beyond maximizing their own earnings (Camerer 2011). For instance, in some contexts people do not like inequality between themselves and others (Fehr and Schmidt 1999; Charness and Rabin 2002). They also care about reciprocating help they receive (Rabin 1993; Kurzban and Houser 2005) and simply helping others without expecting anything in return (Batson et al. 1983).

Is Helping Others Really about Helping Ourselves?

Perhaps the fact that people care about others' outcomes can help them make good climate decisions. Maybe you would vote if you were just as happy to see others receive benefits as you are when you receive benefits yourself. Imagine that you estimate going to the polls will cost you about $20 (for example, maybe you must take time off work to vote), but across the *entire country* people would benefit by a total of $200 million if Candidate Jones is elected. So long as you have slightly greater than a 1×10^{-7} chance of breaking a tie, you should vote ($\$20 = \$200 \text{ million} \times 1 \times 10^{-7}$). A rigorous game theory analysis confirms this intuition (Edlin, Gelman, and Kaplan 2007). If people value outcomes for others, many more people will go out to vote than the very small number who would show up if they only cared about themselves.

Before rushing to conclude that this solves all our problems, we should note that people's motives to do good are not straightforward. People sometimes help just for the "warm glow" that accompanies helping; people feel good helping others, regardless of how effective that help is (Andreoni 1995). In a recent example of this, a team of researchers explicitly informed people about which charities were most effective at delivering aid (Caviola, Schubert, and Nemirow 2020). Nonetheless—and despite recognizing which charities were more effective—many people in this study continued to donate to ineffective charities because they had an emotional attachment to them.

Another issue is that people may be motivated to be nice or to make what seem like nice choices because it makes them look good to themselves or others (Feddersen, Gailmard, and Sanfroni 2009; Caplan 2011). People will even compete over who is the most altruistic (Barclay and Willer 2007). A team of psychologists titled their paper "Going Green to Be Seen" because they found that people chose green consumer products particularly when they wanted to gain status *and* their choices were public (Griskevicius, Tybur, and Van den Bergh 2010). More recent work by this team found that people choosing sustainable products were seen as more attractive as romantic and sexual partners (Palomo-Vélez, Tybur, and van Vugt 2021).

Whether this "competitive altruism" is good or bad in climate politics depends on whether the observable forms of climate aid or mitigation are actually the most useful forms. It's possible that people will forgo more useful ways to mitigate climate change, instead focusing on strategies they can brag about to their friends and neighbors. Or, they may simply not think about whether the "help" they want to provide really does help.

Deciding for Others in the Disaster Game

In our own research, we sought to put all this together. We wanted to test whether people would do a good job making climate decisions for others (see Andrews, Delton, and Kline 2021). So, we turned back to our workhorse game, the disaster game. Recall that in the basic disaster game, everyone starts with a pot of real money. When the game begins, there is a chance that everyone will lose all this money. But if the group can collectively contribute enough to a shared climate threshold, they will prevent disaster and save whatever money they have left.

The first important change we made is that people worked to prevent disaster for *others*. Players were in groups of four, and each player started with 100 tokens worth real money. Of these 100 tokens, 80 could be lost to disaster but could not be spent to prevent disaster, representing illiquid assets such as infrastructure. The remaining 20 tokens could both be lost to disaster and could be spent to prevent disaster, representing liquid assets. Unlike in typical studies, however, groups could not use the spendable money to prevent their own disaster; they could only spend to help a different group prevent disaster. (For any particular group, yet another set of players could contribute to the group's threshold. But, to be clear, there was no reciprocity.) This setup mimics the way that those at risk of disaster often have little control over whether disaster is averted. Again, think about how with climate change it is current generations in rich countries who are making decisions that will mostly affect the world's poor and future generations. In fact, our experiments really amped this up as groups had absolutely no way to stop their own disaster. The basic structure of the game is shown in Figures 10 and 11.

The second change we made was to add more complexity to the basic game. Recall one of the problems facing Kiribati: What type of technology for adaptation should they use—seawalls or mangroves or something else? In our new game, we again asked players to choose between different technological solutions for (simulated) mitigation (see Chapter 3). One solution represented steady, incremental progress, such as wind or solar power. The other represented a risky new idea, but one with a high upside, like geoengineering. Players had three choices. First, they could contribute nothing to the other group. Second, representing incremental technology, players could directly contribute 20 tokens to the other group. Third, representing risky technology, players could make a gamble such that there was a 50% chance that their 20 tokens doubled to 40 and a 50% chance that the tokens disappeared. See Figure 12.

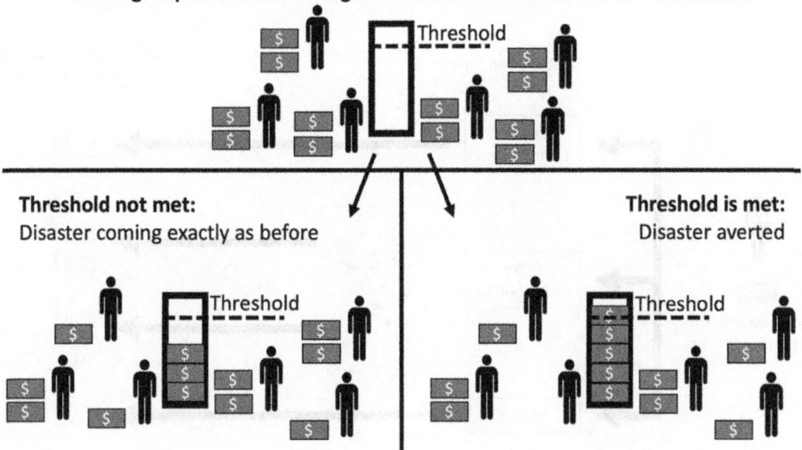

Fig. 10. The disaster game. Everyone starts with personal money (top panel). They each decide how much to contribute to a group threshold. If the combined contributions are less than the threshold, everyone loses their remaining money (bottom left panel). If the combined contributions are more than or equal to the threshold, everyone keeps their remaining money. Contributions disappear regardless of whether the threshold is met.

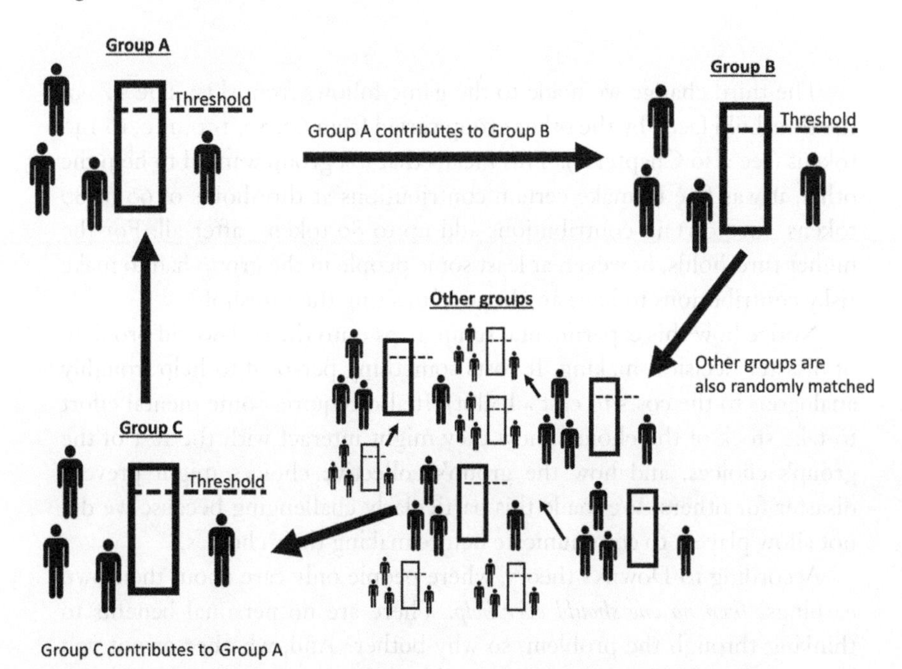

Fig. 11. Among a large set of groups, groups contribute to meet one another's thresholds. However, help is not reciprocal. If Group A contributes for Group B, then Group B does not contribute for Group A. Group B contributes to help a random other group and Group A is helped by a different, random other group.

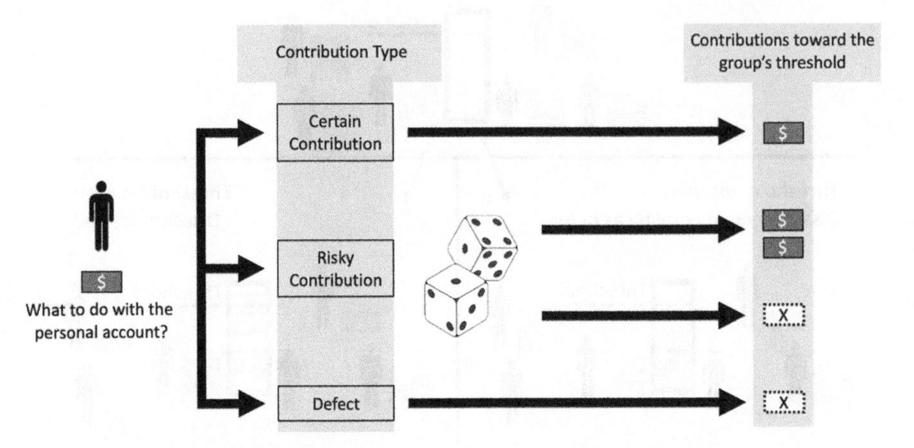

Fig. 12. The decision each player faced. A player starts with a personal pot of money. If they make a "certain contribution," that money is given directly to the group's threshold. If they make a "risky contribution," there is a 50% chance the money will be doubled before going to the group's threshold and a 50% chance the money will disappear and nothing will go to the group's threshold. If they "defect," then they keep their personal money, and nothing goes to the group threshold.

The third change we made to the game follows from this: The size of the threshold faced by the other group varied from 60, 80, 100, 120, to 140 tokens (see also Chapter 3). This means that if a group wanted to help the other, it was fine to make certain contributions at thresholds of 60 or 80 tokens; four certain contributions add up to 80 tokens, after all. For the higher thresholds, however, at least some people in the group had to make risky contributions to have any hope of meeting the threshold.

Notice how this experimental setup maps onto the real-world problem of disaster decision-making. It costs something personal to help (roughly analogous to the costs to cast a ballot). It also requires some mental effort to take stock of the choices, how they might interact with the rest of the group's choices, and how the group's collective choices might prevent disaster for others. We made this particularly challenging because we did not allow players to communicate before making their choices.

According to Downs's theory, where people only care about their own earnings, *then no one should ever help*. There are no personal benefits to thinking through the problem, so why bother? And, whether or not you know what to do, there are no personal benefits to putting that knowledge into action.

If people just want to *look* like they are helping, then, yes, they might be willing to contribute to the other group. But it should not particularly matter to them whether they do a good job with their contributions. (Think back to the people contributing to ineffective charities because they feel emotionally attached to them.) On this view, people might choose randomly between the two contributions. Or they might mostly pick the certain contribution: Although it's not always the most useful, it does guarantee that some help will be received by the other group. If you pick the risky option, then they could get nothing, which might make you look bad.

However, if people are primarily concerned with helping others *in an effective way*, then we should see people mostly choosing to contribute to the other group and doing so in useful ways. Here, "useful" means that direct contributions should be more common at the two low thresholds and risky contributions should be more common at the highest thresholds (as we saw in a similar study covered in Chapter 3).

What do we actually find? We first ran this experiment on a sample of American adults who played the game online, recruited through Amazon's Mechanical Turk, an online convenience sample. When the game began there was a 90% chance that everyone's money would be wiped out. Each of their 100 tokens was worth 1¢ (a stake size typical for this setting). Remember that of these 100 tokens they could spend 20 to help the other group. And spend they did. A whopping 82% of players spent their 20 tokens to help the other group. We stressed to the players that there was no way they could help themselves by spending this money, and we made this abundantly clear in the instructions. We even included a comprehension question to ensure players knew they could not help themselves, only the other group. This 82% figure is only a shade less than what we found in a similar sample of Americans who played to prevent their own disaster, 87% (see Chapter 3). Our players were very willing to help.

This speaks against the simple view that people only care about themselves. But it doesn't show why they want to help. If they want to prevent disaster for the other group, then we should see changes in the rates at which our players choose the certain versus risky option. And that's exactly what we found (see Figure 13). As the figure shows, people were particularly likely to choose certain contributions at the low thresholds of 60 and 80. And they were more likely to choose risky contributions at the higher thresholds of 100 and 120. (Although we were surprised to see it in our data, the dip in risky contributions at the highest threshold, 140, was predicted by our game theory; see Chapter 3.)

With these data, we can see that our players were willing to pay to contribute in some way—an action that is roughly analogous to showing up

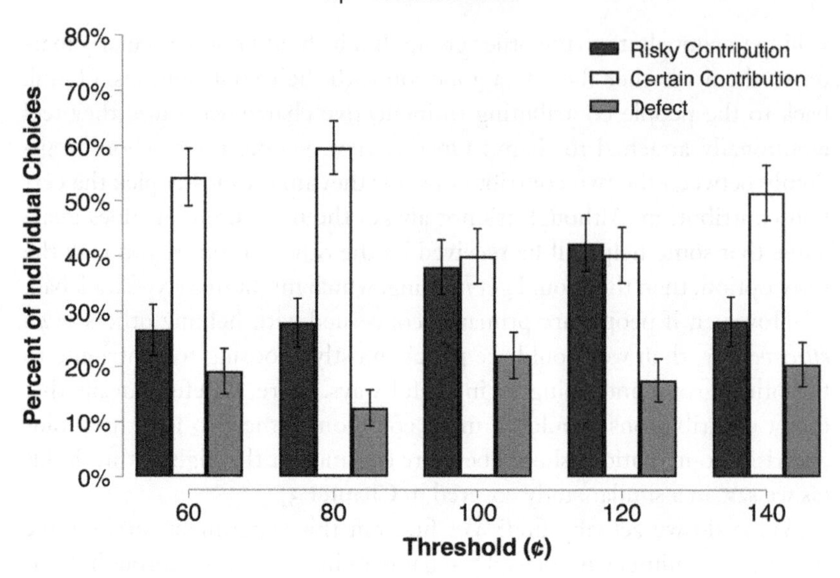

Fig. 13. The percentage making certain, risky, or no contributions at each threshold. Error bars are standard errors of the mean.

at the polls to cast a vote for one candidate or another. But the evidence is less direct that they paid any mental costs to decide which choice was best. They did pick the most useful choices on average, which suggests that they did some thinking about how to help. However, we wanted to provide an explicit way our players could show that they would pay to learn. To do this, we made two changes to our game. First, after players made their initial decision about whether and how to help, we surprised them with an offer: They could buy advice from us on how best to help the other group and then change their original decision if they wanted. If they were willing to pay for advice from us, it would provide direct evidence that our players were willing to spend for information.

Second, because we wanted to make advising them easy, we only used thresholds of 80 and 120 tokens. The most efficient way to solve the problem at the other three thresholds is a bit complex, because they involve different players doing different things. Consider a threshold of 60 tokens. The best way to meet this threshold is for three players to give 20 tokens directly and for the final player to simply do nothing. Or consider a threshold of 140 tokens. The best strategy is for one player to contribute 20 tokens directly and for the others to all choose risky contributions (hoping all three pay off 40). With thresholds of 80 and 120 tokens, however, all four players should do the same thing. For 80 tokens, everyone should

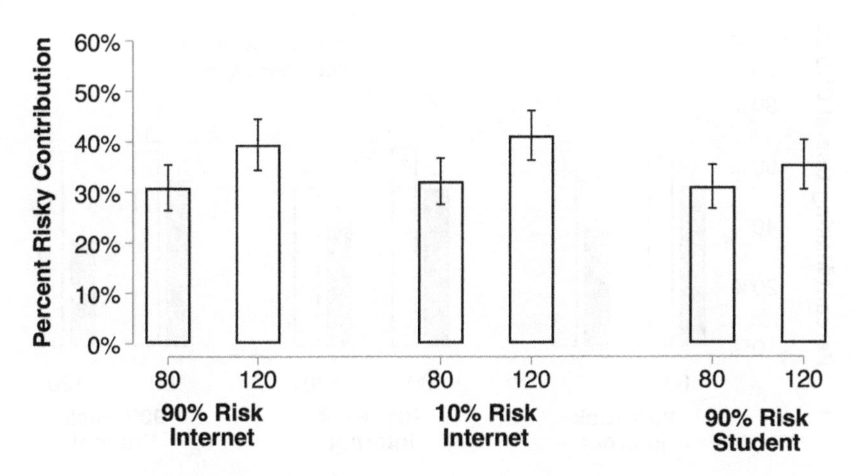

Fig. 14. The percentage making risky contributions at each threshold in each sample. Error bars are standard errors of the mean.

contribute 20 tokens through certain contributions (20 * 4 = 80). For 120 tokens, everyone should make a risky contribution and hope at least three pay off 40 (40 * 3 = 120); the fourth person also making a risky contribution maximizes the chance that enough gambles will pay off.

We ran this experiment twice. Once, as before, we collected an American sample playing online for $1 stakes. The second time we collected a student sample in our own laboratory. Besides giving us more control over the setting (now we could guarantee they weren't also watching TV), this also allowed the game to be played for higher stakes: $20 or $0.20 per token.

First, did we replicate the general finding that players were willing to help? Yes, across both new studies about 87% of players helped the other group, a number *identical* to when people made this decision on their own behalf (see Chapter 3). Second, did we replicate that people made more certain contributions at the low threshold and more risky ones at the high threshold? Yes, as shown in Figure 14, people made more risky contributions at the higher threshold. (Our internet sample was divided into two: One facing the standard 90% chance of disaster and the other a much lower chance of 10%; this change does not affect any results.)

Finally, were people willing to pay for advice from us? Yes, across both new experiments a whopping 72% of players bought information on how best to help. Clearly, many people wanted our advice. For some people, this information merely confirmed the choice they already made. Other players, however, learned that their original choice was not the most helpful. In

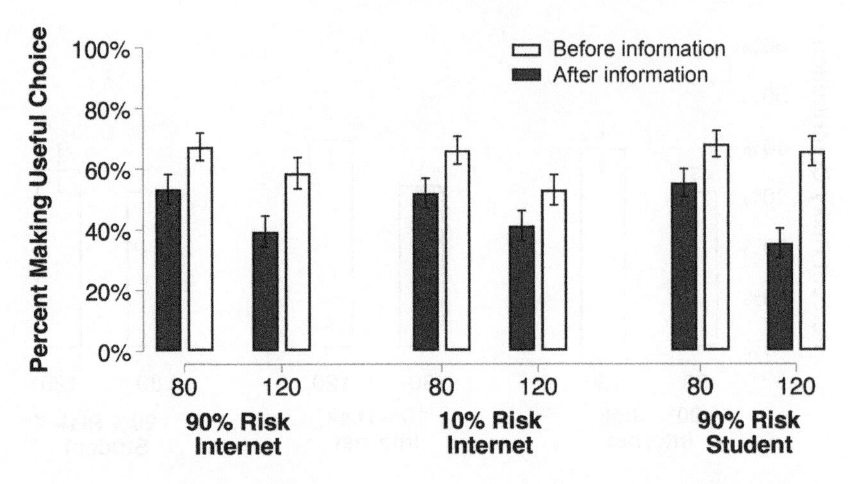

Fig. 15. The percentage of players in each sample making the most useful choice before and after the opportunity to buy information, at each threshold. Error bars are standard errors of the mean.

this set, a bit over 50% changed their choice to match our advice. So, not only did players want to know what to do—and paid for this knowledge—many of them used this information to help others in the best way possible. Figure 15 shows that our samples as a whole made better decisions after the opportunity to buy information. This was true in all three samples and regardless of whether the threshold was 80 or 120 tokens.

In sum, our players can make good disaster decisions for others even when those are costly and complex decisions that offer no potential material benefit to themselves.

Creating Social Preferences to Care about Others

Responding to climate change doesn't always involve sending direct aid to others; instead, it often involves reducing *externalities* (see also Chapter 5). An externality is when one person creates costs or benefits for others, but those costs or benefits are not internalized into the person's own decision-making. For instance, if you do amazing landscaping in your yard, that's a positive externality: Your neighbors benefit from the view, yet this benefit was not why you created the garden—you just like gardening. Another research team modeled negative externalities in the climate (Sherstyuk et al. 2016). Their game simulated the ability of players to emit greenhouse gases (which earned the emitting players money). However, the emissions could build up, causing damage to later generations of players; this latter effect is the negative externality. Unlike in the disaster game, there was no separate mitigation mechanism; instead, players simply had to balance the direct benefits of emissions with the later costs of emissions buildup.

As is the theme of this chapter, the researchers were interested in how people make decisions for others. They wanted to study what happens when climate decisions involve multiple generations of people, with earlier generations affecting later ones (again, see also our Chapter 5). In their game, there's a tension between helping yourself now by emitting (you benefit from your own emissions) and hurting others in the future (your emissions will reduce the payoffs of later generations). In their *separate-generations condition*, new players made decisions every generation, with each generation facing the emissions buildup from earlier ones. If players only cared about their personal returns, they should be particularly willing to emit, and to ignore the consequences they are passing on to others.

They also created two comparison groups. The first and more complicated one we'll call the *early-returns condition*. This condition also has new players every generation. It differs from the separate-generations condition, however, because early players get a return on what happens to later generations (hence our label). Unlike in the separate-generations condition, in the early-returns condition players do not necessarily benefit by emitting as much as they can; emitting too much now hurts not only later generations, but the current players as well. What's interesting about the early-returns condition is that it experimentally creates a "social" preference. Whereas our games simply measured whether people cared about others' payoffs, these researchers used incentives to create players who should care about future generations: Players in the early-returns condition are given more money if they decide well on behalf of others.

The final condition was the *same-players condition*. In this condition, the same players made decisions every generation; there were never new players. Thus, for these players there was no externality. Instead, the game was a type that is called a commons dilemma (which we'll discuss more in Chapter 5). The main challenge was for players to coordinate among themselves to avoid emitting too much. This condition served as a baseline to the two versions with different players each generation. Emphasizing that the early-returns condition created incentives for players to help others, the researchers' game theory showed that players in both the same-players condition and the early-returns condition should play in the same way—they should all focus on balancing the costs and benefits of emissions to maximize earnings across all generations.

Several interesting results pop out. First, in the same-players condition, players emitted about the perfect amount to maximize their earnings over the generations. But in the separate-generations condition, players emitted about 32% more—helping their own generation at the expense of later generations. Nonetheless, these players were not maximally selfish, which would have entailed a rise of 54%. That is, even though players behaved somewhat selfishly, they were also somewhat willing to help.

Turning to the early-returns condition, where current players' earnings were tied to future players', things were even rosier: Now they only emitted about 12% more than what would have maximized everyone's earnings. In other words, as with our studies, players made good decisions for others. Again, the benefit of this experiment is that social preferences are experimentally created: Players are incentivized to care about later generations.

The researchers also allowed players to give advice to the next generation. Not surprisingly, in the same-players condition, the players emphasized maximizing their own payoffs by being circumspect in their emissions. These players focused on how everyone is better off if people were careful not to emit too much. In the separate-generations condition, however, the most common advice was to think about what was best for oneself. In the early-returns condition, where there were incentives to care, players were about split between advice to focus on their own payoffs and advice to think of others' payoffs.

Continuing with the advice theme, another research team studied an intergenerational disaster game with *nudging* (Böhm, Gürerk, and Lauer 2020). Nudging refers to small, nonbinding interventions that might help people to make better decisions (Thaler and Sunstein 2009). In this game, each generation had the option to invest in two different funds as well as to contribute to a disaster threshold. One fund paid a greater interest rate than the other. But investing in the lower return fund gave the *next* generation more money to help them meet the threshold. Thus, there was an intergenerational tension: Get paid yourself or help the other group more.

This research team nudged by making the default investment decision the amount that would be best for everyone across all generations. Players, however, could easily override this default. As it turned out, this nudge did lead to greater investments in the fund that helped the next generation.

The researchers also created an experimental manipulation that made people care about future generations. They asked players to make a pledge that they would not "place the interest of my own group above the interests of subsequent groups" and that they would "act in solidarity toward the members of subsequent groups, so that they do not have to face conditions that are worse than the conditions for my group" (p. 8). After making this pledge, people behaved more nicely, taking into consideration the future generations. In other words, pledging to be generous increased generosity.

Altogether the game evidence shows that people can make complex decisions for others, including ones that involve preventing disaster. Our games studied this using the desires to help that people brought with them into the lab. We showed that people will evaluate complex problems and arrive at good decisions. Other teams successfully used experiments to manipulate concern for others. This jibes with other sources of evidence from outside the lab showing that people can prepare for real disasters in effective ways to help others (Lindell and Perry 2012). As an example, some research shows that people can tell when disasters are likely to be especially bad and will evacuate accordingly (Dash and Gladwin 2007). And people in real disasters will seek out and integrate information, at least when that information comes from trusted sources (Lindell and Perry 2012). (What happens if sources are not trusted? See our Chapter 6.)

Making Dangerous Decisions for Others

There are many ways that people can make climate decisions, including ones that affect others. Governments of large countries can unilaterally impose a carbon tax or other regulations on their citizens. Consortiums of nongovernmental actors can invest in technology accelerators. Private citizens can make choices about what sort of products they buy, in turn affecting the decisions of businesses and corporations. Our next focus is climate technology that falls under the umbrella of "geoengineering."

Geoengineering covers a lot of ground (Keith 2000). What sets it apart from other technologies is that geoengineering's unofficial motto is "go big or go home." For example, an incremental technology would be to make more efficient cars that emit less CO_2 per gallon of gasoline. Geoengineering would include negative emission technologies, like a machine that can pull large amounts of CO_2 from the air and store it safely. An incremental technology would be more efficient solar cells and better batteries. A geoengineering technology would be to seed large swaths of the atmosphere with aerosols, to reflect more sunlight back into space.

Precisely because they are large and intentional efforts to modify the environment, geoengineering is under intense scrutiny. Scientists, policymakers, and journalists have been debating not only the merits of individual geoengineering technologies but whether we should be discussing them at all. There are two main issues (Keith 2000). One is the problem of "moral hazard"—the worry that even just discussing a moon-shot geoengineering technology will crowd out thinking and investment in necessary incremental technology; we'll return to this in Chapter 6. The other problem is that geoengineering a complex system like the climate could have unintended consequences.

The back and forth over geoengineering was illustrated by the public debate series Intelligence Squared. Two pairs of experts debated the motion: "Engineering solar radiation is a crazy idea" (Intelligence Squared 2019). In the debate, the energy analyst Anjali Viswamohanan likened deploying geoengineering technology to the wacky misadventures of the animated sci-fi duo Rick and Morty—full of unintended, dangerous mayhem. There are huge potential downsides with geoengineering; Viswamohanan suggests that geoengineering could disrupt rain cycles, oceans, and the ozone layer.

Clive Hamilton, Viswamohanan's debating partner, raised the possibility of bad actors using the technology to enrich themselves or their country at the expense of others. Do we want the Chinese government using this technology at the expense of Indians? Do we want the Kremlin to control it? The scientist David Keith, debating here on the pro-geoengineering side, had earlier written that Soviets and Americans began studying geoengineering as a military strategy as part of the Cold War (Keith 2000).

Perhaps more worrisome, some geoengineering tech is so cheap that even individuals could get involved (see Chapter 1). A man once dumped more than 100 tons of iron dust into the ocean off Canada (Falconer 2018). The idea was that the iron dust would stimulate the growth of plankton and the plankton would in turn absorb CO_2. His actions went against two United Nations treaties and got him labeled a "rogue geoengineer" and an "eco-terrorist."

Despite these possible downsides, geoengineering might still be necessary. As we noted in Chapter 3, the IPCC argues that at least some uncertain technologies, especially negative emissions technologies like carbon capture, will be needed (IPCC 2021). Keith, in his debate remarks, pointed out that cutting or even eliminating emissions does not *solve* the problem, it only stops *adding* to it, because the effects of emissions last for so long. In his view, even pulling carbon from the air will not be enough; solar radiation modification will be required. Keith's co-debater, Ted Parsons, adds that a coherent policy requires considering all possible tools, even ones that seem crazy at first. (If you're curious, the pro-geoengineering side won the debate by a wide margin.)

A Geoengineering Game

We designed an experiment to understand how people think about geoengineering. Specifically, we want to know how people respond when there is a chance that geoengineering will backfire, and when the person deciding whether to implement geoengineering is safe from those consequences. Keep in mind that there is little geoengineering tech available at wide scale yet, so this experiment was forward looking: a thought experiment about a future problem brought to life in the lab.

We again turned to the disaster game. This time, however, we kept the basic game simple: There was a single threshold of 60 tokens and each player in a four-person group could spend any amount they wanted from a 30-token bank account. Whatever was spent was given directly to the threshold. As before, the contributions represented incremental action. Each player also had a larger pot of 70 tokens, representing nonliquid assets such as buildings and roads. When the game started, it was 100% certain that disaster would cause all four players to lose 35 tokens from this large pot. But if everyone pulled their equal share and contributed 15 tokens, they would prevent disaster. Preventing disaster saves each player a net 20 tokens, a huge portion of their total stake.

The novel feature here is that players could use (simulated) geoengineering. Consistent with the very low costs of some geoengineering, we made using geoengineering free. If they used geoengineering and it worked, then the four players didn't have to do anything—disaster was completely averted, and they were able to keep all their money. However, if geoengineering was unsuccessful, it *backfired*. In this case, the four players each lost 35 tokens from their 70-token pot, *and* they still faced the possibility of disaster. In other words, besides losing 35 tokens from geoengineering backfiring, they stood to lose an additional 35 tokens if they did not contribute enough to the threshold to stop climate change (see Figure 16). Of course, if geoengineering was not used, then it could not backfire, but the group definitely faced the original problem. When the group decided whether to contribute to the threshold, the players were told both whether geoengineering was used and whether it worked.

We experimentally manipulated two features of this game. First, we manipulated who was given the power to geoengineer. In the *control condi-*

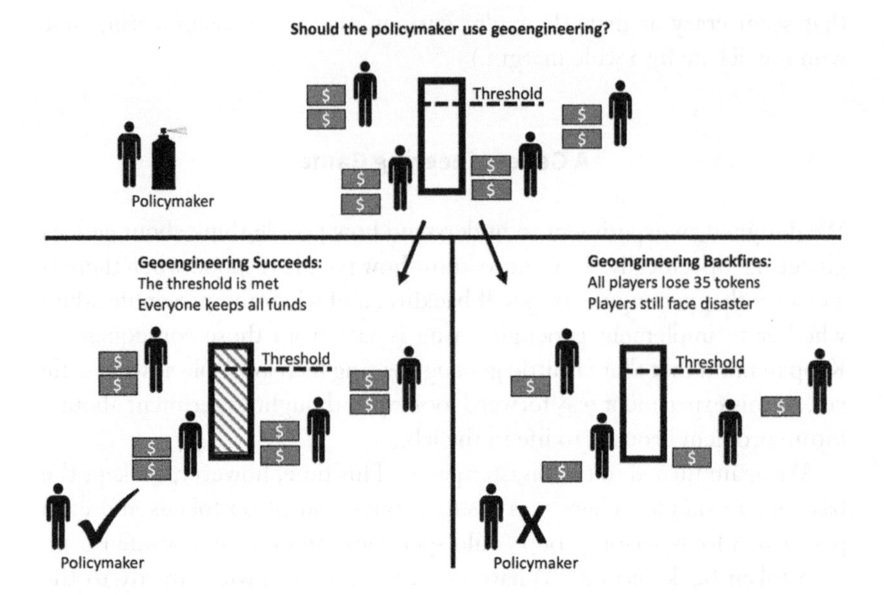

Fig. 16. The basics of the backfire game. Four citizens each independently decide how much to give to a threshold to prevent disaster. A policymaker decides whether to enact geoengineering. If the policymaker uses geoengineering, it could succeed and meet the group's threshold. But if geoengineering fails, it also backfires. When it backfires, citizens lose money and continue to face the same challenge of preventing disaster.

tion, one member of the four-player group was randomly selected as the group's policymaker, and this person unilaterally made the geoengineering decision. Importantly, the policymaker had skin in the game: If they used geoengineering and it worked, they avoided having to spend any money to stop disaster; if it failed, they lost money from it backfiring, just like everyone else.

In the *outside actor condition*, a fifth person was part of the experiment. This fifth person was the policymaker and, as before, could unilaterally decide whether to use geoengineering. Like the other players, the outside policymaker was given a total stake of 100 tokens. However—and this is the key change—the outside policymaker suffered no ill effects if geoengineering backfired or from the original disaster if it happened. They were *guaranteed* to keep all their money no matter what. This means they had no skin in the game.

Second, we manipulated how likely it was that geoengineering would succeed, with possibilities of 1%, 20%, 40%, 60%, 80%, and 99%. To be clear, if geoengineering did not succeed, then it definitely backfired; so, the complement of these probabilities represents the chance of backfiring. The technical appendix to this chapter has a game theory analysis showing that at low probabilities geoengineering should never be used and at high probabilities it should always be used. At low probabilities of success, the risk of backfire is too large, and the group is better off using incremental contributions. The exact cut point depends on whether or not the policymaker believes the group will meet the threshold through incremental contributions. If the policymaker thinks the group *will* meet the threshold without geoengineering, then the policymaker should only use geoengineering when it is really likely to succeed—80% and 99%. But, if the policymaker thinks the group will *not* meet the threshold, then they should also use geoengineering when it has a 60% chance of success. The intuition for all this is straightforward: If the group is likely to reach the incremental threshold, there is no reason to expose them to unnecessary risk. We weren't sure if real players would be sensitive to this subtlety, but we looked for it in our data.

Notice that this setup differs from our previous games in this chapter. Earlier, *only* the outside players could help. In the backfire game, however, the outside policymaker could abstain from using geoengineering and leave the problem to the group itself. Moreover, in the previous games, the outside decision makers will—at worst—leave things at the status quo. But with the current game, the policymaker's decision could actively *hurt* the group.

This design allows us to examine three questions. First, will people use geoengineering less when it has a higher chance of backfiring? Presumably, greater risk will lead to less geoengineering; this straightforward finding would serve as a sanity check for the rest of the experiment. Second, and more interesting, how will outside policymakers' choices compare to those made by group members? Perhaps outside policymakers will use geoengineering in exactly the way insiders do. Or perhaps their tolerance for risk will differ from the group's own. Third, if there is such a difference, does it depend on how likely geoengineering is to backfire? The results for these last two questions help us understand how real policymakers might make decisions about geoengineering, as those decisions will primarily affect other people.

Risky Decisions for Others

A typical study on risky decisions might work as follows (see also Chapter 2). Research participants must decide between receiving $10 for sure or a

50% chance of receiving $30. If the person cares only about the expected payoffs, then they should pick the gamble: 50% of $30 equals $15, which is greater than $10. But many people are risk-averse, meaning they prefer not to shoulder too much risk even if that means receiving less overall. A risk-averse person might prefer the sure $10 over a gamble for $30.

Looking beyond decisions for the self, researchers have also studied how people make risky decisions for others. In studies like this, the question is: Compared to deciding for themselves, will people make riskier or less risky choices when deciding for others? Research finds that people will pay to reduce the amount of risk they or others face, suggesting that people do not want risk and believe others do not either (Fornasari, Ploner, and Soraperra 2020).

But if you must make a risky decision for others, what do you do? According to a recent review, the answer is: It depends (Polman and Wu 2019). Overall, these researchers found a small tendency for people to make riskier choices when deciding for others. But this depended on who the other person was: When the other person was a child or a medical patient, people became more risk-averse; when the other person was a friend, a family member, or, especially, a stranger, people made riskier choices. They also found that when the studies emphasized gains *or* losses, people were more likely to make risky choices (see Chapter 2 on gains and losses).

The upshot is that it's unclear from past work what to expect in our experiment. On the one hand, our players are relative strangers to one another, and the game is clearly about loss prevention. This suggests that we should find a shift to risky choices when deciding for others (i.e., more geoengineering by outside policymakers than group members themselves). On the other hand, past work was not usually about groups or thresholds for disaster. Perhaps such a strong disaster context will evoke the same mindset that people use when thinking about medical patients. This would make people take fewer risks. Finally, many studies simply find no difference between deciding for oneself or others (Polman and Wu 2019). Perhaps our outside policymakers will decide as if they were members of the group.

Geoengineering Decisions When Backfiring Is Possible

First, we examined how likely players were to use geoengineering based on how likely it was to succeed. As shown in Figure 17, players clearly used

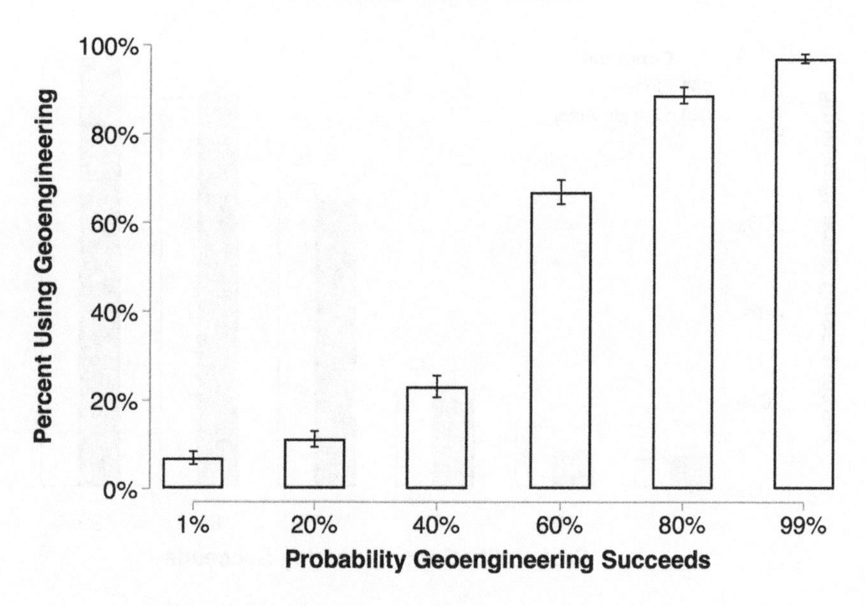

Fig. 17. The proportion using geoengineering at each probability that geoengineering succeeds. Error bars are standard errors of the mean.

more geoengineering when it was more likely to succeed, and therefore when it was less likely to backfire. The game theory predicts that it should be all or nothing and the real data came pretty close to this: At the three lowest probabilities of success, few people used geoengineering; at the two highest, many used it. This gives us confidence that whatever else we find is not an artifact of people failing to understand the game.

Turning to the critical question, let's see whether players decided differently when deciding for others. As shown in Figure 18, it turns out that they did not. At all probabilities of success, people were just as likely to use geoengineering whether they were outsiders or group members. Our outside policymakers played the game as the group would have wanted. Keep in mind that at no point did we suggest to outsiders that they should try to do what the group wanted; it was entirely up to them how to play.

As in our previous studies, players appeared to care about a group they did not belong to. Although using geoengineering was free, it took some mental effort to follow the instructions and to decide what to do based on the probability of success. Outside policymakers were clearly willing to pay these mental costs. Had they been unwilling, they would have chosen randomly.

To our surprise, policymakers in both conditions were sensitive to the subtlety we noted above: If you think the group *can* meet the threshold

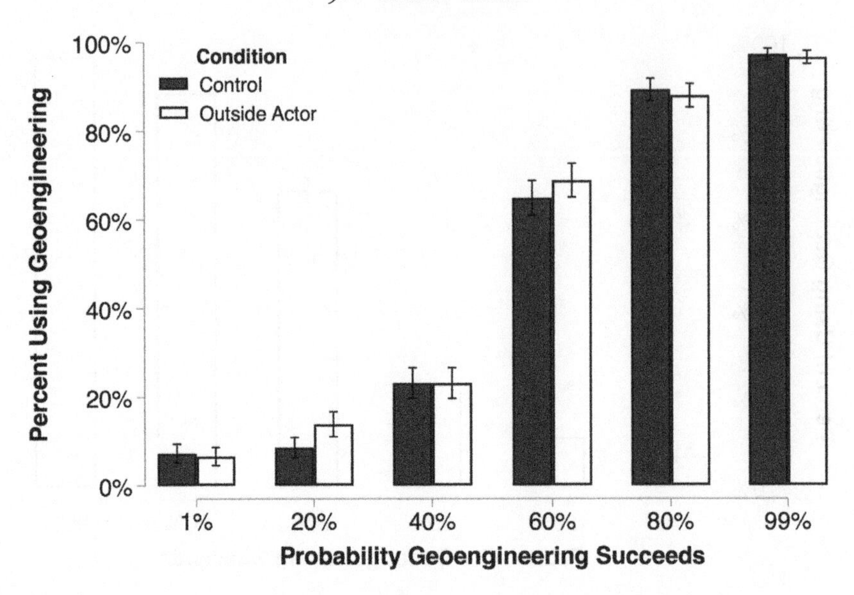

Fig. 18. The proportion who used geoengineering at each probability that geoengineering succeeded, in each condition. Error bars are standard errors of the mean.

without geoengineering, you should only use geoengineering at 80% and 99% chances of success. If you think they *cannot* meet the threshold on their own, you should also use geoengineering at a 60% chance of success. As shown in Figure 19, players followed this pattern: At most probabilities *except* 60% (and to a lesser extent 40%), policymakers were just as likely to use geoengineering regardless of what they expected the group to do. But *at* 60% (and to a lesser extent 40%), policymakers were more likely to use geoengineering if they thought the group could not meet the threshold on their own. Yet again, our players surprised us with their ability to pick up on game theory nuances.

The upshot of this study is positive: When outside policymakers must decide whether to use moonshot technology that could backfire, they could be overly risk seeking if they face no consequences. Or, they could be overly risk averse if they are worried about being culpable. In fact, outside policymakers were calibrated to what the group itself wanted.

Of course, this does not imply that all real-world policymakers will make good decisions. Real policymakers face other incentives that bias their decisions one way or another. A politician from a rich country who is deciding on a policy that would affect a developing country must consider

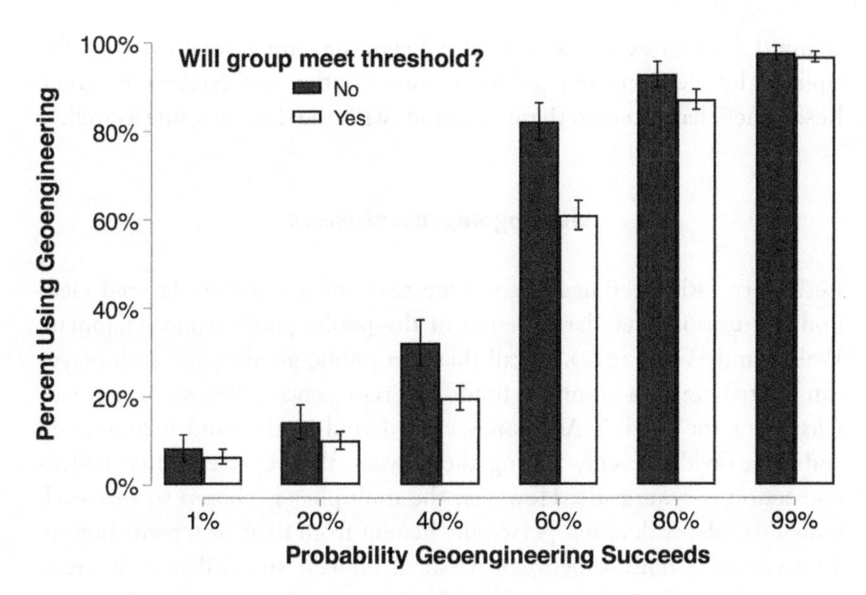

Fig. 19. The proportion who used geoengineering at each probability that geoengineering succeeded. Results are broken down by whether players believed their group would succeed without geoengineering. Error bars are standard errors of the mean.

how her choices will be interpreted by her *own* constituents, not just how her choices impact the other country. If her constituents have a bias toward action (Sunderrajan and Albarracín 2021), this may push her to take too many risks. Nonetheless, our results do show that there is nothing intrinsic to the process of making decisions for others about dangerous disaster that would lead policymakers astray.

Asymmetry in the Ability to Decide

We have mostly considered very stark situations of deciding for others: In the above games, there was a clear distinction between the people making the decision and the people affected. In our games where people could avert disaster for others, players could not choose who was helping them or communicate to ask for help. A similar setup was used in the backfire game when the policymaker was an outsider.

But often, the people who are affected have some influence over the people making the decisions. For instance, poor and developing countries participate in international meetings that produce principles on climate action (Chandler et al. 2002). Real policymakers are at least partially disciplined by elections and public opinion (Butler and Nickerson 2011). Researchers have studied these situations with the disaster game as well.

Electing Representatives

Let's start with electing leaders. One team of researchers layered elections on top of a standard version of the public goods game (Hamman, Weber, and Woon 2011). Recall that in a public goods game each player can contribute to a common fund; this represents collective action (see Chapter 2 for details). All money contributed to the fund is multiplied and then divided evenly among the players; this represents that collective action is synergistic. However, the multiplier is chosen so that each individual player does not personally benefit from their own contribution. This creates a tension between personal interest and collective interest. For instance, if in a group of four players the multiplier is 1.4, then each $1 contributed becomes $1.4, and divided four ways this is only $0.35. So, if you gave $1, you would only get $0.35 back from your effort. Players following their own self-interest should not bother contributing anything. On the other hand, if all players contribute $1, then everyone gets $1.4—a 40% rate of return.

In typical versions of this game that allow a group to play multiple rounds—contribute, see what happened, contribute again—players often begin by contributing a substantial amount but lower their contributions as the rounds progress. This research team replicated this common finding. Their players started by contributing about 40% of their stake. But after a few rounds this trended down to about 20%. Past work has shown that allowing players to communicate with each other can help prevent this slide. The researchers found this, too. When they allowed players to communicate, contributions jumped up to about 90%. This was short-lived, though, and eventually contributions slid back down to 20%.

What happens when players can elect someone from their group to make decisions for them? Can contributions to the public good be sustained? The answer was a resounding yes. When players could elect someone to make decisions for everyone in the group, the elected leader chose contributions for everyone that were up around 90%, just like when communication was allowed. Unlike under communication, however, elected leaders sustained this high level of contributions throughout the entire game. All players did exceedingly well when they elected a leader, much better than in the standard version of the game or when they had non-binding conversations. When players selected their leaders, they seemed mostly to avoid people who would, as individuals, contribute little to the group fund.

What happens in the disaster game with elections? Things might be different because public goods are about creating a positive benefit, not preventing a loss. Luckily, a research team headed by the creators of the disaster game addressed this question with an ingenious design (Milinski et al. 2016). Their *election condition* involved eighteen players divided into six groups of three players each. Each group elected a representative and the six representatives played a multi-round version of the disaster game. The results of this game affected the earnings of all eighteen players. That is, each elected official chose how their group as a whole would contribute to the threshold, rather than each individual deciding for themselves. This version was made to map onto international negotiations between elected representatives.

The researchers contrasted this election condition with two control conditions: a *six-person control* and an *eighteen-person control*. In these conditions, all players made decisions as individuals about whether to contribute to the threshold. To keep things constant, the threshold was three times as large in the eighteen-person condition compared to the six-person condition. What's beautiful about this design is that the control conditions cover two important comparisons. The six-person control matches the number of representatives in the election condition; the eighteen-person control matches the number of total people in the election condition. This way, the researchers can tell if it is uniquely the election, or just differences in the number of people, that causes any differences they might find in contribution behavior.

So, are representatives better than individuals at preventing disaster? In the first stage of the experiment, the representatives were not actually elected. Instead, they were randomly chosen by the experimenters and participated in a multi-round disaster game. These random representatives did not do a great job. The six of them prevented disaster only 33% of the time, the same as the eighteen-person condition where everyone made decisions on their own. The small groups in the six-person control were much more successful; they prevented disaster 60% of the time. Not surprisingly, meeting the target is easier when there are fewer people who need to coordinate. Unfortunately, randomly selecting representatives from a large set of groups does not seem to help.

Of course, real representatives are not usually thrust into their role by pure happenstance. In the election condition, after the first stage of the experiment, elections were held and the chosen representatives played another multi-round disaster game. In this way, players could vote out representatives who didn't do what the group wanted and could retain good representatives. After this, yet another election was held and yet another multi-round disaster game was played. In the two control conditions, the individuals simply played two more multi-round games.

In the eighteen-person control condition the players succeed more as the game went on, their success rising from the original 33% to hover around 50%. Players in the election condition also did much better, with their success rising from 33% to nearly 70%. This reveals the key result: When the total number of players is held constant, electing representatives helps prevent disaster. However, small groups did even better: As the experiment progressed, the six-player control ended with about 90% of groups avoiding disaster. In other words, small is better, but if that's impossible—which is the realistic case—elected representatives can help.

Groups in this experiment tended to elect hard bargainers (Milinski et al. 2016). Many representatives walked a tightrope such that the threshold was met yet their group got away with contributing less than others. This kind of aggressiveness might not work in all strategic encounters, but it was effective here. When one person seems committed to making only small contributions, everyone else is better off making up the difference and preventing disaster.

Another study found a different, but related, drawback to electing representatives (İriş, Lee, and Tavoni 2019). In this study, representatives heard from their constituents about what they wanted. Unfortunately, the representatives mostly listened to the stingiest of their constituents, which caused them to contribute less to the threshold than they otherwise would have. One would prefer that representatives listen to everyone, not a biased sample.

The Rich and the Poor

In the first studies in this chapter, we idealized the process of making decisions for others. The people affected by disaster cannot help themselves in any way; an entirely different set of people makes all the decisions. In the real world, things are not usually this stark. Rich countries have considerable resources they can invest in mitigation. Poorer countries have less, but not none. Developing countries face considerable risks from climate change. Developed countries face fewer risks, but again not none. Thus, real climate decisions involve some actors with *more* decision-making power or *less* to lose, not *unilateral* power or *nothing* to lose.

How well do people prevent disaster when they vary in their risks and endowments? Several studies have used the disaster game to answer this question, mostly focusing on varying endowments. Unfortunately, the evidence is mixed. The first study, by Alessandro Tavoni and colleagues, hewed closely to the original disaster game (Tavoni et al. 2011). Players working in groups of six had to contribute, over ten rounds, to a threshold of €120. If they failed, there was a 50% chance they would lose their earnings. Each player began the game with €40. However, for the first three rounds, the players did not decide for themselves how much to contribute. Instead, the researchers had a computer program make decisions for each player. In the control condition, the computer allocated €2 from each player to the climate threshold; notice that if all six players contribute €2 every round, then they would meet the threshold with everyone contributing their equal share (€120 = 6 players * €2 * 10 rounds). In the inequality condition, for each of the first three rounds, the computer allocated €4 from three players and €0 from the other three players. This created "poor" and "rich" players, respectively. Notice that in both conditions the computer forces each *group* as a whole to contribute identical amounts to the threshold, but only in the inequality condition were *individual* players forced to contribute different amounts. After these initial three rounds, players made their own contribution decisions for the remaining seven rounds.

Inequality made preventing disaster harder. In the control condition, 50% of groups were able to prevent disaster. In the inequality condition, however, only 20% of groups did so. This seemed to be because rich players were often unwilling to contribute extra during later rounds to offset the early costs paid by poor players. Thus, even though these rich players had more power to help prevent disaster, they proved unwilling to flex this power on behalf of others.

On the positive side, Tavoni and colleagues created additional conditions where they allowed players to make nonbinding pledges about their own future contributions. When players could make pledges, they were better at stopping disaster. Without inequality and with pledges, 70% of groups escaped disaster (up from 50% in the original control condition). In an inequality version with pledges, 50% did so (up from 20%). Note that in this study players were poor *because they had already contributed a lot to mitigating disaster*, a feature that does not clearly map onto real-world emissions and mitigation.

Another early study by the team who created the disaster game also examined the effects of inequality, with a slightly different design (Milinski, Röhl, and Marotzke 2011). Unlike in their original disaster game, players in this new game had one pot of money they could spend on disaster and another they could not spend but could lose if disaster happened. (This is similar to our studies on choosing between different technologies; see the first studies in this chapter and Chapter 3.) As before, the threshold was €120 for six players. If not met, there was a 90% probability of disaster. This study had three conditions: all poor players, all rich players, or an equal mix of both. Poor players had spendable pots of €20 and non-spendable pots of €30; rich players had pots of €40 and €60, respectively. Notice that even when all players were poor, they could reach the threshold, though this required spending all the money they could (6 players * €20 = €120).

Here, inequality did *not* undermine group success. Mixed groups and rich groups invested the same amount, generally about as much as would be needed to meet the threshold. (Poor groups were not as successful, possibly because they would have to spend quite a lot, as a relative percentage of their pots, to prevent disaster.)

Why the difference between these two studies? One possibility is that Milinski and colleagues used a 90% chance of disaster, but Tavoni and colleagues only a 50% chance. A greater disaster probability might have increased willingness to contribute. Another possibility is the origin of the inequality between players: Milinski and colleagues randomly assigned players to be rich or poor; there was no connection between wealth and past contributions. This pure randomness might have encouraged contributions from rich players (see Kameda et al. 2002).

A later study by Max Burton-Chellew and colleagues created variations in both wealth and risk (Burton-Chellew, May, and West 2013). In this study, the group had to reach a threshold of 120 tokens (each token was worth £0.50). A group of six was given 240 tokens they could contribute to meet the threshold. For egalitarian groups, this was divided equally as 40 tokens per player. For unequal groups, poor players had 20 tokens each and rich players 80 tokens each. The novel twist in this game is that, within a group, some players faced greater risks of disaster than other players. (This was true for unequal groups. Players in egalitarian groups always had identical risks.) In some unequal groups, rich players faced a greater risk of unmitigated disaster. In other unequal groups, poor players faced a greater risk. And, as a control, sometimes both types faced the same risk. The researchers argue that the version where poor players face more risk than rich players is most consistent with the real world: The richest countries, who would be best positioned to spend on mitigation, are also less at risk of major climate disaster.

This condition that was most consistent with the real world was also the most disastrous: Disaster was rarely prevented for unequal groups when it was the poor who were most at risk. Only one of eight such groups (12.5%) staved off disaster. There were no differences between the other conditions; in all other cases, 75% of groups prevented disaster. This converges with Milinski and colleagues' study, which also found success from unequal groups. When rich and poor groups had the same risk or the rich had the greater risk, the rich were willing to spend *proportionally* more than the poor. But when the poor were especially at risk, the rich were only willing to spend the same proportion as the poor, and this wasn't enough to solve the problem.

Another study, by Thomas Brown and Stephan Kroll (2017), examined whether unequal groups were successful when they faced a threshold of uncertain size and when there was an uncertain risk of disaster if the threshold was not met. (Here the risk of disaster, whatever it might be, was the same for all players.) Compared to complete certainty, uncertainty in the size of the threshold lowered contributions substantially and prevented all groups from meeting the threshold. On the other hand, compared to complete certainty, uncertainty in the risk of disaster only lowered contributions a bit and depressed success somewhat. But the key finding for our purposes here is that inequality did not matter: Groups of unequal players and of equal players contributed similar amounts.

In sum, across these four studies, the pattern is mixed. Three studies find that inequality is not particularly problematic in most cases (Milinski, Röhl, and Marotzke 2011; Brown and Kroll 2017; Burton-Chellew, May, and West 2013). One study finds that inequality hurts success (Tavoni et al. 2011); this is paralleled by a study of our own that was modelled after the Tavoni study, which we discuss in Chapter 5. A more recent study finds evidence both ways (Vicens et al. 2018). In this recent study, *all* groups succeeded, whether players were equal or not. However, the contributions appeared intuitively unfair: Poor players often contributed more than richer players.

On balance the results are positive, but show reason for concern. When players face differing incentives (e.g., when the poor face greater risk; Burton-Chellew, May, and West 2013) or when there is the potential for disagreement about who is responsible (as in Tavoni et al. 2011; Kline et al. 2018), success is unlikely. This is perhaps why the players in our studies at the beginning of this chapter were so generous: It was absolutely clear who did and who did not have the power to prevent disaster; in those studies only others had the power. This is why clear roles and values are important for solving collective problems. This can be seen in conventions about shared but differentiated responsibilities for mitigation, a topic we turn to in the next chapter.

When Others Are at Stake

In sum, we found largely positive evidence that people are good at deciding for others. In our study testing whether people could solve disaster for others, players spent to help and made good decisions even though the problem was complex. In our backfire game, outsiders who were immune to disaster made choices exactly as insiders did. Other work finds that elected representatives effectively managed collective decisions, including in the disaster game. The only rough spot was when there was inequality in both assets and risks. Inequality alone was not fatal, but when the rich simultaneously had less exposure to risk, they did not help enough to solve the problem for the poor.

Technical Appendix

This appendix provides more details about the experiment on geoengineering backfiring that was reported in the main body of Chapter 4.

Additional Methods

The players were 302 people recruited though Amazon Mechanical Turk, an online convenience sample that completes surveys and other tasks for small amounts of money. Our sample was only US adults who had not participated in any of our other disaster games. They received $0.50 just for participating and started with a bonus pot of $1. The game was played only once, and players played asynchronously. So, although we matched players into groups for payoffs, all players were statistically independent for the purposes of analysis.

We manipulated between groups whether the policymaker was a part of the four-person group or an independent fifth player. In other words, players only experienced one version of this manipulation and did not know the other manipulation existed.

The rest of the study used the "strategy method." The strategy method asks players to make each possible decision they *could* face, without knowing which one they *actually* face (i.e., without knowing which decision determines their payment). Thus, in the control condition, all players were asked to decide what they would do as the policymaker. They were shown each possible probability that geoengineering would succeed and asked for each probability that they would use geoengineering. After deciding whether to use geoengineering, they then made decisions about contributing to the threshold, deciding as if they were not chosen as the policymaker. They contributed once assuming that geoengineering was not used, and once assuming that it was but had backfired. (Remember that if geoengineering is successful, no contributions are necessary.) Although each player knew what their own geoengineering decision was, their decision was not guaranteed to be the one used for their group. Thus, even players who used geoengineering could sensibly be asked how much they would contribute if geoengineering was not used.

In the outside actor condition, all players were first asked to assume that they were the outside actor. They then decided, for each possible chance that geoengineering could backfire, whether they would use geoengineering. After their geoengineering decisions, they were asked to assume instead that they were part of the four-person group and to make contribution decisions. As before, they did this once assuming that geoengineering was not used, and once assuming that it was but had backfired. Note that in both conditions, when players were asked to contribute in the case of backfire, they were not aware of what the initial probability of backfiring was, only that it had happened.

Game Theory

Here we describe the game theory of what agents should do if they want to maximize expected payoffs for the group. The table shows the parameters and the values for them that we used in the experiment.

First, in the absence of geoengineering, the payoff-dominant equilibrium is to contribute. To show this, we check whether the value of each person contributing their fair share is greater than or equal to the value of everyone defecting.

TABLE 5. Definition of each parameter

Parameter		Value in Game
Endowment	E	70 tokens
Personal Account	P	30 tokens
Fair-Share Contribution	C	15 tokens
Threshold	T	60 tokens
Disaster Penalty	D	35 tokens
Geoengineering Penalty	G	35 tokens
Probability of Geoengineering Success	X	{manipulated}

$$E + P - C \geq E + P - D$$

Because the disaster penalty, *D* (35 tokens), is greater than the fair-share contribution, *C* (15 tokens), in the absence of geoengineering the payoff-dominant equilibrium is cooperation.

The next question is: At what probability of geoengineering success, *X*, should the policymaker implement geoengineering? This answer depends on what they think the group will do if the group does have to decide whether to contribute to the threshold (i.e., when geoengineering is not used or when it is used but fails). In both instances, they should use geoengineering when the expected value of doing so is greater than the expected value of not doing so.

If the policymaker believes the group will *fail* to meet the threshold, then they should use geoengineering when:

$$X(E + P) + (1 - X)(E + P - D - G) \geq E + P - D$$

This equation reduces to:

$$X > G/(D + G)$$

This means the probability of geoengineering's success has to be greater than the backfire penalty relative to the sum of both penalties. Given our game parameters, if the policymaker believes the group will not meet the threshold, then the policymaker should implement geoengineering when the probability that it will succeed is greater than 50%.

However, if the leader believes the group will *succeed*, they should use geoengineering when:

$$X(E + P) + (1 - X)(E + P - C - G) \geq E + P - C$$

This reduces to:

$$X > G/(C + G)$$

This means *X* must be greater than the size of the backfire penalty compared to the size of the backfire penalty plus the cost of disaster prevention. In our game, this means *X* must be greater than 70%. In other words, the probability of geoengineering backfiring must be less than 30% for policymakers to use it if they believe the group will contribute to the threshold successfully.

Putting together these two thresholds—50% and 70%—explains why in the main chapter we checked at a 60% chance of success whether policymakers choose differently based on their beliefs about the group's ability to meet the threshold. A chance of 60% is the only probability we studied that falls between 50% and 70%.

Results

Because we had no reason to think the relationship between the probability of geoengineering success and the rates of using it would be a straight line, we analyzed this relationship with a linear probability model with a separate dummy variable for each probability and used the 1% chance of success as the suppressed reference group. As shown in the table, there was an effect for each dummy variable. Across conditions, people were more likely to use geoengineering when it was more likely to succeed (see Figure 17 in the chapter).

TABLE 6. Results of a linear probability model regressing whether the policymaker used geoengineering on the probability that geoengineering succeeded. The omitted base category is a 1% chance of success. Standard errors clustered at the level of the respondent in parentheses.

Probability Geoengineering Succeeds	(1)
20%	0.04***(0.01)
40%	0.16***(0.02)
60%	0.60***(0.03)
80%	0.82***(0.02)
99%	0.90***(0.02)
Constant	0.07***(0.01)
N	1812

* $p < 0.05$, ** $p < 0.01$, *** $p < 0.001$.

To examine whether outsiders used geoengineering more, we used an independent-samples *t*-test to compare the control condition to the outside-actor condition, separately for each probability of success (see Figure 18 in the chapter). As shown in the table, the tests were not significant for any comparison. Altogether this suggests that it did not matter whether decisions were made by outsiders or insiders.

We also checked whether there was a difference in choosing geoengineering depending on whether the policymaker did or did not believe the

TABLE 7. *T*-tests determining whether there are differences across conditions in whether policymakers use geoengineering. The first column shows the probability that geoengineering succeeds. The second shows the proportion using geoengineering when deciding for their own group, while the third shows the proportion using geoengineering when deciding for others.

	Proportion Using Geoengineering			
Pr(Success)	Deciding for Self	Deciding for Others	*t*-statistic	*p*
1%	0.07	0.07	0.23	0.82
20%	0.09	0.14	1.46	0.15
40%	0.23	0.23	0	1
60%	0.65	0.69	0.73	0.46
80%	0.89	0.88	0.36	0.72
99%	0.97	0.97	0.34	0.74

TABLE 8. *T*-tests determining if there are differences in whether policymakers use geoengineering if they believe the group will meet the threshold without it. The first column shows the probability that geoengineering succeeds. The second shows the proportion using geoengineering when they believe the group will meet the threshold without geoengineering, and the third shows the proportion using geoengineering when they believe the group will fail without it.

	Will the group meet the threshold without geoengineering?			
Pr(Success)	No	Yes	*t*-statistic	*p*
1%	0.08	0.06	0.58	0.56
20%	0.14	0.10	1.03	0.30
40%	0.32	0.20	2.30	0.02
60%	0.82	0.61	3.56	0.0004
80%	0.93	0.87	1.40	0.16
99%	0.98	0.97	0.38	0.71

group would meet the threshold. The game theory shows that policymakers have a tighter window for using geoengineering if they think the group will be successful with incremental contributions. In fact, we did find this difference, as shown in Figure 19. As shown in the table below, *t*-tests find no differences between rates of using geoengineering for four of the six probabilities of success. The only exceptions are at 40% and, especially, 60%. Players who thought the group would not incrementally succeed were more likely to use geoengineering at these thresholds. Keep in mind that we did not experimentally manipulate policymakers' beliefs, so this is not an experimental comparison.

Although our main interest was in the policymakers' behavior, we also summarize the contribution decisions of the groups. Players contributed 19 tokens on average, 63% of their stake, an amount clearly over the 50% needed from each player to meet the threshold. There was no difference in contributions depending on whether geoengineering was not used or was used but backfired ($t(301) = -0.83$, $p = 0.41$).

FIVE

Flirtin' with (Self-Created) Disaster

Many disasters are outside of our control; a volcano erupts on its own schedule. Others we inflict on ourselves intentionally; wars damage by design. Climate change is different because we contributed to it without meaning to; it's an accidental self-created disaster.

The basic problem is that the atmosphere can only absorb so much greenhouse gas before the earth warms too much, with *too much* usually defined as 1.5°C or 2°C above preindustrial levels. The amount of greenhouse gas that can be emitted before we reach this point is called the *carbon budget*. Staying below the carbon budget wouldn't be difficult if greenhouse gases quickly cycled out of the atmosphere. With quick cycling, the carbon budget would quickly renew, like how most people's budget renews with a paycheck every week. Unfortunately, greenhouse gases do not cycle quickly. Greenhouse gasses like methane remain in the atmosphere for around twelve years, while the more abundant carbon dioxide remains for centuries. Because the carbon budget renews only slowly, historical emissions matter at least as much as current emissions. The emissions of earlier generations remain for us, and ours will remain for later generations. It's like getting the same amount as your weekly paycheck, but only once a year.

To make matters more complicated, not everyone has emitted equally. China tops the global chart for current emissions, but its per capita emissions are well below those of other high emitters like the United States and Canada (see Figure 20). Other industrial nations follow close behind. Developing countries are contributing a growing, though historically modest, share. Critically, historical emissions and wealth are closely related:

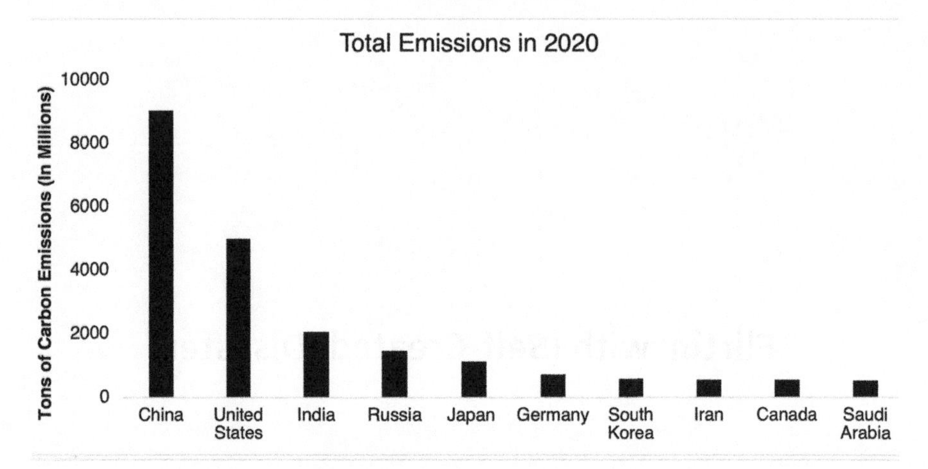

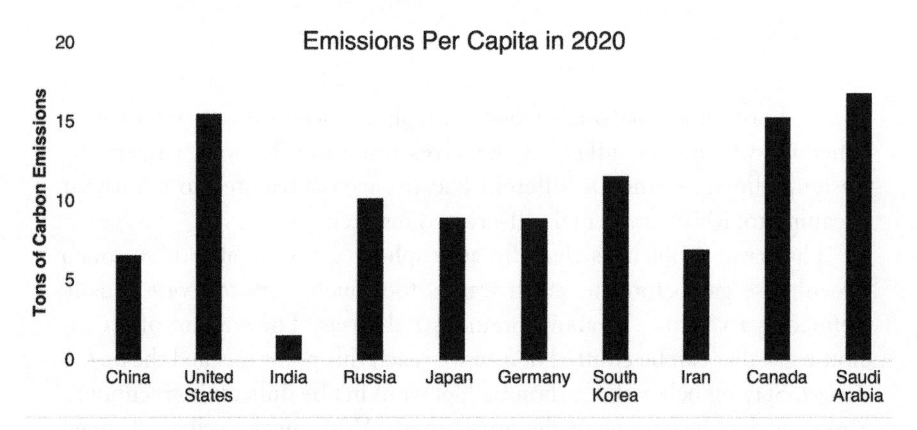

Fig. 20. Total carbon emissions and carbon emissions per capita for the top ten highest-emitting countries in 2020.

Rich countries are rich *because* they emitted or benefited from others' emissions. Being rich, historically high emitters are in the best position to spend on disaster prevention and adaptation. These facts are why, in treaties and negotiations over climate change, rich countries are generally expected to shoulder greater obligations than poor ones.

This chapter looks at a series of related questions: Can people avoid creating disaster? How is this affected by differences in wealth? Can people solve social dilemmas that they themselves created?

Taking from the Commons

To recap, there's a limited carbon budget, which is being used up as countries emit greenhouse gasses. If we go past this budget, we face climate disaster. Game researchers have been studying dilemmas like this for decades, called *commons dilemmas* (or often also called *common pool resource dilemmas*). A commons is a resource that many people can use because it is communally owned, or not owned by anyone. Imagine an ocean fishery in international waters; this is a commons that anyone can harvest from. The more fish you catch, the more you earn. The problem is that there must be enough fish left to reproduce and replenish the stock. If a lone fisher harvests, no problem; one person can only take so much. But if many fishers harvest, they may deplete the stock beyond what is sustainable. Wouldn't the mass of fishers realize this and not overfish? After all, everyone is better off if everyone shows restraint than if everyone harvests all they can. If everyone shows some restraint, taking only a reasonable amount, everyone gets to fish indefinitely. But if everyone fishes too much, the fishery quickly collapses.

The problem is that restraint is good for fishers as a group, but not for fishers as individuals. If everyone else is restraining themselves, then one fisher can fish all they want without destroying the fishery. They might as well take as much as their boat can haul. Of course, if everyone thinks this way, then everyone will overharvest and the fishery will collapse. (There are parallels with this dilemma and the prisoners' dilemma and public goods game discussed in Chapter 2.)

It's much the same with the carbon budget. If only one person in the world was emitting carbon dioxide, it wouldn't really matter. But when billions of people are emitting, it adds up quickly. Similarly, each nation emitting just a little bit of carbon dioxide is not a big problem, but when many large nations are emitting millions of tons of carbon dioxide each year it is catastrophic. And, again, restraint is not individually rational: My carbon emissions are just a small drop in the metaphorical bucket, so why would I bother reducing my emissions? Carbon emissions represent a commons dilemma.

We are not the first to describe the atmosphere as a commons (Dietz, Ostrom, and Stern 2003). Just like other commons—fisheries, forests, and aquifers, for example—the carbon budget has a couple of key technical properties (Ostrom, Gardner, and Walker 1994). First, commons are *not excludable*. When something is not excludable, it is difficult or impossible to prevent others from using it. Turning back to the fishing example, how can one lone fisher keep someone miles and miles away from also catching the fish? Second, commons are *rivalrous*. This means when one person uses it, there is less for others. If I catch this fish, then you cannot eat it. True, in the long term the stock might replenish because more fish will hatch. Although the atmospheric commons replenishes itself, the long-lasting nature of carbon dioxide means that it does so over a very long timescale. We do not always have the luxury of waiting for the long term—as the economist John Maynard Keynes said, in the long run we are all dead.

The terminology can get a bit confusing, and we will not get too hung up on terms—we will always describe the experiments to make clear what properties the games have. Nonetheless, it will be helpful to compare commons dilemmas and public goods a bit more (see Chapter 2 on public goods). Briefly, public goods are when a group of people can collectively produce something useful but each individual is better off letting others do the production. The textbook example is national defense. The entire nation is better off with an effective army. But each individual citizen would be even better off if they could avoid paying the taxes for it: Because their taxes are a miniscule percentage of the military budget, they'll still be protected even if they find a way not to pay.

Sometimes commons are contrasted with public goods. Neither a commons nor a public good is excludable—you cannot prevent others from using either. But only a commons is rivalrous. For example, a lighthouse would be a public good in this sense: Two boat captains can see the light without detracting from the other's view. As another example, a levee is a public good: It protects an entire neighborhood, and one home being protected doesn't prevent another home from avoiding flood damages. Sometimes, though, researchers use public good in a broader sense, focusing only on excludability and ignoring rivalry. Under this definition, a commons is just a type of public good.

It's also important to keep in mind that both rivalry and excludability aren't either-or distinctions. In most cases we can think of goods as being more or less excludable and more or less rivalrous. Take the lighthouse. If the light is on, all ships can see it, so it comes pretty close to being perfectly nonexcludable. It is also nearly perfectly nonrivalrous because one ship seeing it doesn't diminish its visibility for any other ships. Similarly, mitigation of climate change is almost perfectly nonrivalrous: It doesn't matter who actually does the mitigation or where it takes place, the benefits of additional mitigation are not diminished. This is why many call it a *global* public good. Consider another example, a canal. A canal is fairly easy to make excludable; you just need to set up a toll booth. And it's moderately rivalrous. Up to a point, additional boats do not diminish others' use of the canal, but eventually congestion causes rivalry as boats get in one another's way. (The same is true for roads.)

Let's look at a representative study that uses a game to model a commons dilemma. It will illustrate features that the atmospheric commons shares with other commons, and how it differs from the public good of climate change mitigation. Games modeling commons dilemmas map well onto the problem of *emissions*. In real-world commons like fisheries, there is a fixed budget of fish that can be sustainably harvested; in the atmospheric commons there is a fixed budget of greenhouse gases that can be sustainably emitted. The game we describe next is therefore very different from a game like the disaster game. Disaster games map better onto the problem of *mitigation*: Contributions need to be made to prevent worse damages later, but everyone would individually prefer not to make them. (Later in this chapter, we'll see games of our own that combine both these features.)

One of the most influential researchers to study commons dilemmas was the late Elinor Ostrom. Her work was so influential that, in 2009, she won the Nobel Prize in Economics, becoming the first woman to do so. She achieved this distinction despite not being officially trained as an economist (her PhD was in political science). Ostrom's pioneering work combined case studies of how people govern real-world commons and economic games. Following her innovative multimethod research, other scholars have taken advantage of the portability of economic games, exporting commons games from the laboratory and into people's everyday environments. Economic games allow researchers to see how real people who make a living from the commons—like fishers, farmers, and foresters—respond to changes in their environment, whether those changes reflect ecological changes, changes in laws, or changes in informal expectations. Many game experiments (including our own) involve students as participants. Games out in the field, however, broaden the subject base by studying working adults from all over the world.

In economic games, researchers typically try to make the payoffs *salient* to the participants (see Chapter 2 for more on salience). To make payoffs salient, they should cover players' opportunity costs, that is, what they could have earned from another activity during the time they participated in the experiment. For the typical experimental laboratory with undergraduate student subjects, this usually means paying what a student could earn at a work-study job, or something around minimum wage. When commons experiments recruit participants whose livelihood depends on the use of a commons, researchers often calibrate the payments so that they match what the participants might earn on a typical day in their work. When decisions are made in a familiar context with similar stakes, the findings probably map more closely onto what the players would do in their real lives.

One such experiment with professionals is a study of abalone fishers in Mexico (Finkbeiner et al. 2018). Abalone are a type of sea snail, similar to a clam. The researchers recruited 180 abalone fishers from six cooperatives in the Baja California region in northern Mexico. They participated in experiments that reflected the common-pool fishing from which they earn their livelihood. They were randomly placed in groups of five. Each session began with a stock of 100 (hypothetical) abalone as the group's common-pool resource. In each of 15 rounds every fisher could harvest a maximum of 5 abalone, and they earned 15 Mexican pesos per abalone they harvested. Given this, their potential earnings matched what they could make from a typical day of fishing. After each round, the members of the group learned how many total abalone were harvested, but they were not told how many abalone each person took. After harvesting, the experimenter created the amount of abalone in the next round by increasing what was left in the common pool by 10%.

Can the fishers make the abalone last until the end of the game? In commons games, the ease of doing so is determined by two things. First is the *stock*: how much of the resource is currently available. In this study, the amount was 100 abalone at the outset of the experiment. The second thing to matter is the *flow*: the number of additional resources added to the stock during each time period. In this experiment, it was 10% of the remaining abalone. Given the specific numbers for this game, it turns out that if all members of the group harvest the maximum, the stock will be exhausted before the fifth round, only one-third of the way through the game. Prudence is therefore necessary to sustainably harvest the resource and to continue to earn money, just as in their real-life occupations.

There are many ways that uncertainty creeps into commons dilemmas. One source is social: The outcome depends on the actions of all the group members. There can also be uncertainty about the rate at which the resource regenerates, which the researchers captured in their experiment. The version of the abalone experiment described above, with a certain regeneration of 10%, was only the researchers' baseline version. They also created several variations. First, they incorporated environmental uncertainty through a one-in-ten chance of a "mass mortality" event after each round. This was determined with a 10-sided die, which was rolled in view of the entire group. If a 10 came up, 50% of the remaining abalone stock was destroyed.

In a second variation of the game, the researchers introduced a "poacher." Despite many real-world fisheries being notionally restricted to a particular set of people, it can be difficult to exclude all outsiders from common pools, perhaps especially fisheries. (And the atmosphere!) So, after each round in the poacher version, a 10-sided die was rolled to determine how many abalone the poacher took from the pool. The die was rolled out of sight of the participants—the fishers knew how much of the stock was depleted each round, but they couldn't know how much of the harvesting was done by the other members of their group and how much, if any, by the poacher. (Recall that in no version of the game are players told how much each member of their group harvested; they only learned the total amount harvested each round.)

In a third variation, participants were told that the "government" had granted them exclusive rights to their traditional fishing grounds. Though the government allowed them to harvest up to 5 abalone per round (as in the baseline condition), they were told that the cooperative had democratically agreed to a quota of just 2 abalone per round. These fishers were, in their workaday world, organized into democratically governed cooperatives, and their government had granted these cooperatives exclusive fishing rights, so this feature of the game was intended to mimic another aspect of the participants' real occupational environment. However, as in the baseline condition, participants did not know how much each other player took from the fishery. So, while they were told that everyone agreed to only take 2 abalone per round, they could still take up to 5 with impunity.

Finally, the researchers introduced conditions where the players could communicate with each other before deciding how many abalone to take each round. They introduced communication in some of the baseline runs, as well as some of the mass-mortality runs.

In the baseline condition, groups started off at high levels of harvesting, taking about 18 abalone per round. This amount gradually decreased as the resource became scarcer over the course of the experiment, with average groups harvesting approximately 5 abalone per round by the final round. In the baseline condition, when players could communicate, they only took about 14 abalone per round at the outset. Rates of harvesting were also relatively stable, fluctuating between 10 and 14 per round. So, overall, the resource was better maintained with communication. Perhaps surprisingly, the possibility of mass mortality—half the stock dying unexpectedly—did not change by much how the players appropriated. However, when the possibility of mass mortality was combined with the ability to communicate, the participants were able to work together to keep resource use at sustainable levels, stabilizing total group harvest at around 5 abalone per round beginning in the fourth round.

When the participants were told that there was a democratic agreement to limit harvesting to 2 abalone per round per person, groups were able to keep their overall harvest to the voluntary quota of 2 per person or approximately 10 for the group. Thus, they were able to sustain the resource significantly better than in the baseline treatment. When the poacher was added, participants increased their cooperation, reducing their harvesting even more.

These results show that commons with uncertainty can be successfully managed by groups, so long as they are able to communicate among themselves or have appropriate norms of resource extraction. We think it's particularly notable that the 2-abalone quota was simply imposed by the researchers, not created by the players themselves. We suspect that self-created norms would work even better. Notice, too, that there was no way to enforce the norm, yet it still worked. Presumably, an enforcement mechanism would also help the norm reduce the amount of abalone people took per round (Ostrom, Walker, and Gardner 1992).

Communication also helps people cooperate in the disaster game (see Chapters 3 and 4). Indeed, researchers repeatedly find that communication helps people solve social dilemmas (Ostrom 2015). Following Ostrom, many studies of the commons have focused on how people establish and enforce rules, which communication can facilitate. For instance, how do people divvy up the costs and benefits of maintaining the commons? How do they punish those who flout the rules of the commons? Ostrom stressed that successful rules must be tailored to the particular people and the particular environment. In fishing communities, for example, users regulate who can fish where, what types and sizes of fish can be harvested, and when fishing can happen at all. In the experiment, when faced with a potential mass extinction event, participants who could communicate managed to come up with effective agreements to harvest from the commons sustainably.

What kinds of rules do people come up with to ensure that commons are used sustainably? Researchers have found that one marker of success is that users of the commons see the rules as simple and flowing from some notion of fairness—transparency, equality, proportionality, or something else that is intuitive, given the particular commons at stake (Becker and

Ostrom 2003; Cox, Arnold, and Villamayor Tomás 2010). For the climate, the most influential principle of fairness is called *common but differentiated responsibilities*. Let's see how it's tailored to the context of climate change, and in particular, the fact that climate change is self-created.

From Each According to Their Responsibilities

Most negotiations over the climate have been conducted under the auspices of the United Nations Framework Convention on Climate Change. (This is usually called the UNFCCC, but we'll call it the *UN Climate Framework*.) Negotiators recognize that economic development and emissions are intertwined. This is where common but differentiated responsibilities come in: Responsibilities are common because all people will be affected by climate change. Responsibilities are differentiated because certain countries have emitted the lion's share of historic greenhouse gasses.

Since climate negotiations began in the 1990s, the principle of common but differentiated responsibilities has been a focus and a sticking point. The principle was first introduced into international law as Principle 7 of the Rio Declaration on the Environment from 1992 (Pauw et al. 2014):

> States shall co-operate in a spirit of global partnership to conserve, protect and restore the health and integrity of the Earth's ecosystem. In view of the *different contributions* to global environmental degradation, States have *common but differentiated responsibilities*. The developed countries acknowledge the responsibility that they bear in the international pursuit of sustainable development in view of the pressures their societies place on the global environment and of the technologies and financial resources they command. (Emphasis ours)

Consistent with the self-created nature of the climate disaster, this formulation links emissions and other pollution ("different contributions to global environmental degradation") to economic development ("the technologies and financial resources they command"). The principle is reiterated in the UN Climate Framework, also in 1992:

> The Parties should protect the climate system for the benefit of present and future generations of humankind, on the basis of equity and in accordance with their *common but differentiated responsibili-*

> *ties* and *respective capabilities*. Accordingly, the *developed country Parties should take the lead* in combating climate change and the adverse effects thereof. (Emphasis ours)

This formulation is sparser, but two things stand out. First, it adds "respective capabilities" to the principle, making more explicit the link between economic development and responsibility for mitigation. Second, it calls for developed countries to take the lead in combating climate change. While there is wide agreement on the principle, some countries have resisted putting it into practice.

Consider the Kyoto Protocol of 1997. According to the Protocol, industrialized countries like the United States, most of Europe, and Japan were given greater responsibility for mitigation precisely because they were industrialized. Despite the intuitive appeal of differentiated responsibilities, the Kyoto Protocol failed because at least one important actor—the United States—did not accept the division of the mitigation burden. Under the Protocol, the United States would have been obligated to substantially reduce its emissions. At the same time, developing countries like China, India, and Indonesia, because of their low historical emissions, would not have been so obligated.

Though the Kyoto Protocol was unsuccessful, the United Nations did not abandon its goal of differentiated responsibilities. In 2015, the UN Climate Framework facilitated the Paris Agreement, which 192 countries have signed to date. Each signatory made a voluntary commitment to fighting climate change. In this case, a commitment is a public promise to help fight climate change primarily by reducing their emissions to a certain, precommitted level. The plans typically include several distinct efforts, such as the adoption of clean energy, forestation, electric vehicles, energy efficiency upgrades, and more. However, unlike the Kyoto Protocol, the Paris Agreement is silent about exactly how countries should differentiate responsibilities or what those responsibilities should be. With no formal metrics, countries define their own responsibilities. Although it remains to be seen whether countries will follow through, they are at least giving lip service to the principle. For instance, when India submitted its commitment to mitigation, it stated, "The principle of common but differentiated responsibilities is the bedrock of our collective enterprise. . . . By enhancing their efforts in keeping with historical responsibility, the developed and resource rich countries could reduce the burden of their action from being borne by developing countries" (UNFCC 2015). When it is described this way, we suspect that more readers will see common but differentiated responsibilities as an intuitive norm, particularly for a self-created problem like climate change. Nonetheless, precisely because anthropogenic climate change is self-created, the norm has remained a contentious issue in international climate negotiations.

How good are real people at implementing common but differentiated responsibilities in situations where their disasters are self-created? To what extent do they agree on *how* responsibilities should be differentiated? We designed a series of games to find out.

Self-Created Disasters

Along with our colleagues Nick Seltzer, Evgeniya Lukinova, and Autumn Bynum, we created a version of the disaster game that allowed us to study whether and when people will create their own disasters (Kline et al. 2018). Before getting to the details, here are the main features. The game is divided into two phases. The first is an economic development phase. In this phase, all players have access to a commons that is designed to represent the carbon budget. The more a player harvests from the commons, the more the player earns. The second phase is the disaster phase. It is like the standard disaster game in that players need to work together to meet a threshold or else they lose everything. The twist, though, is that the development phase and the disaster phase are connected: The more that players, as a group, harvest from the commons in the development phase, the higher the threshold to prevent disaster and the greater the probability of disaster if they fail to meet the threshold. Exploiting the commons creates disasters that are more challenging. Will players restrain themselves? Will they be able to prevent whatever disaster they do create?

Because players individually decide how much to take from the commons, this game allows some players to become richer than others. This feature was inspired by the experiments from Chapter 4 by Alessandro Tavoni and colleagues (2011). The question in those experiments was whether wealthier players would shoulder extra responsibility. As it turned out, they did not; groups with inequality did not do a great job at preventing disaster. However, these players were randomly allocated their wealth; in our games, wealth differences are created by the players themselves. Will inequality, when it's self-created, still prevent players from averting disaster?

The Game

Players worked in groups of 6 people and there were 12 groups in the experiment. The experiment started with the economic development phase. In each of 10 rounds of development, a player could harvest between $0 and $4 from the commons. Thus, players could harvest a maximum of $40. (Harvesting was the only way players could earn money from the game.) This harvesting represents economic growth by emitting greenhouse gases. After each round, they learned how much everyone else had harvested. Harvesting itself was not rivalrous; every player, if they wanted, could harvest the maximum $40 (thus, the group could combine for $240). But, as we'll see in a bit, rivalry cropped up elsewhere in a more subtle way.

In the next phase, the disaster phase, players were given a threshold. If they did not meet the threshold, then there was a chance that they would face disaster and lose all their money. Players had 10 rounds to contribute to the threshold, and in each round they could contribute anywhere from $0 to $4. Their only money was what they harvested during the development phase. After each round, players learned how much everyone else had contributed.

If this was a typical disaster game, there would be no connection between how much they harvested and how difficult it was to prevent disaster. In such a setup, any sane player should harvest as much as possible every round, because they would earn more money to take home and to help meet the threshold. But the catch in this game is that the severity of the disaster was self-created: The more players harvested, the bigger the threshold *and* the higher the risk of disaster if players failed to meet the threshold. Players knew before the development phase started how the two phases were connected. For each dollar they harvested, the group's threshold would increase by 53¢. (Nothing hinges on the exact amount, and 53¢ is admittedly a bit arbitrary.) The more they harvested, the worse the problem.

You might think this alone would make players rein in their harvesting, taking less than the full $240. But that's not guaranteed. If a group collectively harvests just $100, the threshold will be $53. Exactly meeting the threshold leaves the group $47, a bit less than $8 per player. What happens if the group harvests all $240? The threshold will be $127.20. Meeting this leaves the group $112, nearly $19 per player. This hypothetical group is better off harvesting everything and then contributing enough to meet the massive threshold they created. That's why we also had harvests determine the *risk* of unmitigated disaster, otherwise the setup would not be rivalrous like a real commons.

In most disaster games, there is a fixed risk of disaster for groups that do not meet the threshold (e.g., 90%). Here, though, we tied the risk to the group's own harvest. Whereas the threshold rose continuously with every dollar harvested, the risk of disaster rose in a few discrete jumps. The discrete jumps were intended to model tipping points in the climate system (see our discussion in Chapter 3). If players harvested one-fourth or less of the total possible commons, there was a one in six chance of disaster—just 17%. If they harvested between one-fourth and one-half, the risk of disaster was 1 in 2—50%. Between one-half and three-fourths harvested, the risk of disaster was 3 in 4—75%. And for more than three-fourths harvested, the risk of disaster was 11 in 12—a huge 92%. We chose these particular numbers because, when a group failed to meet the threshold, we rolled a real 12-sided die in front of them to determine whether they escaped disaster.

Again, the exact numbers are a bit arbitrary; the important point is that more harvesting means a higher chance of disaster. The increasing risk of disaster transformed the economic development phase into a commons dilemma. First, it was not excludable: Any player could benefit from harvesting. Second, harvests were rivalrous in that each dollar harvested by one player is a dollar that could not be harvested by another player *without running the risk of making disaster more likely*. This final feature has the potential to make players think twice about pushing their harvests to the max. A group that maxed out its harvests would have lots of money, but they would also face a big risk of losing it all.

Did players restrain themselves when harvesting? At first glance, not really: On average, each player harvested $31.30. This is pretty close to the maximum possible of $40. (If everyone in a group did this, it would expose the group to the highest possible risk of disaster, a nearly certain 92% risk.) On the other hand, players did not literally extract as much as they could have. Plus, at least some players told us they were motivated by restraint. After the game, we asked players to write down why they made the choices they did. One of them wrote: "I tried to appropriate [i.e., harvest] enough to keep below a certain amount according to what the rest of the group was appropriating. I wanted to aim for 120 max, but then it went over quite quickly, so I tried to aim for below 180, but unfortunately, [that] did not happen."

Nonetheless, did players contribute enough to stop the disasters they created? Recall that however much money the group had in hand after the development phase, in the disaster phase they need to contribute 53% of their funds to guarantee that they kept the rest of their money. Amazingly, the average contribution rate was only a *single percentage point* lower than

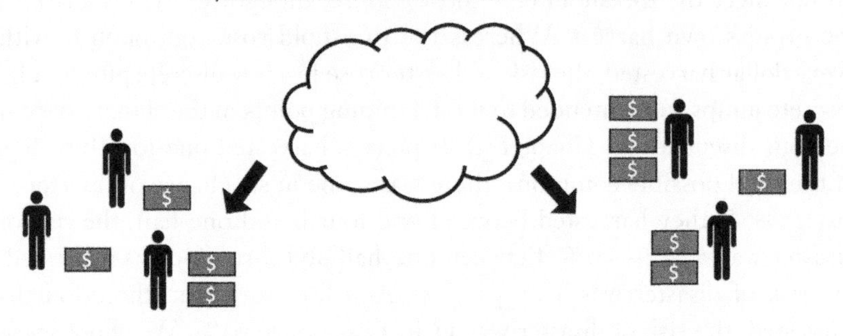

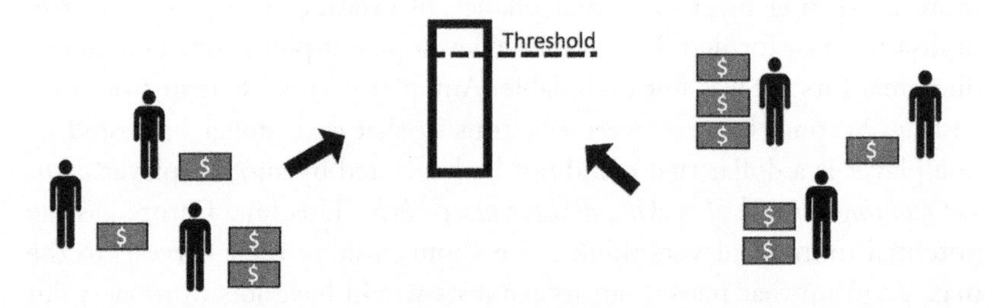

Fig. 21. The two phases of the self-created disaster game. The top panel shows the harvest phase, where each of the six players decides how much money to take from the commons. The bottom panel shows the disaster phase, where the same players have to decide how much of the harvested funds to contribute to prevent disaster. The cost of disaster prevention and the risk of disaster are determined by how much players harvested.

this, 52%. But thresholds are unforgiving: Barely missing the threshold is as bad as not even trying. (In fact, it's even worse: Contributed money is definitely gone; kept money might survive disaster even when the threshold is not met.) Despite the high contributions, only just over half the groups, 7 of 12, contributed enough to meet the threshold.

Did many groups fail because they created their own disasters, or because some players were wealthy and some were poor when the disaster phase started? Recall that players individually decided how much to harvest, so some were rich and others poor when the disaster phase started. That is, some harvested a lot, raised the disaster threshold, and had a lot of funds. Others harvested very little, helping keep the threshold low, but also had fewer funds to help prevent disaster. In the study we saw in the last chapter, compared to groups with equal players, groups with inequality often failed to prevent disaster (Tavoni et al. 2011). Perhaps it's inequality and not self-creation that hindered our groups from meeting the threshold.

To see which is a better explanation, we conducted a second study that included only the disaster phase. In the new version, no one harvested; players simply played the disaster game over 10 rounds. We created 12 groups for the new study. Each new group was matched with one from the original study. Each new group was given the threshold size and risk of disaster of their matched old group, and each new player was given the cash harvested by an old player from their matched group. (We did not tell new players that this was how their group's numbers were created.) In sum, we have our original, self-created condition and our new condition, which we'll call the placebo condition. The disaster phases of both conditions were identical except that players in the placebo condition did not create their threshold, risk, or starting cash. This means that the self-created groups and placebo groups had the same *amount* of inequality, but only for the self-created group was this their *own doing*. It turns out that this difference matters: Placebo groups contributed, on average, 54% of their cash to the threshold. Though that sounds like a tiny increase from the 52% contributed by the self-created groups, it was enough to ensure that *all 12* placebo groups met the threshold. Self-creation caused players to contribute too little to avert disaster, even when holding inequality constant.

History Matters

Our studies and those of Tavoni and colleagues (Tavoni et al. 2011) tell us something about how people react to inequality and disaster, but there's a complexity they're missing: How should developing countries catch up economically? This has also been a perennial concern of the UN Climate Framework. Industrialized nations developed in part through centuries of emissions. For many reasons, developing countries missed that opportunity. If developing countries drastically increased their emissions, they could eventually catch up in wealth. This would require that developing countries eat up big chunks of the carbon budget. To avoid going over budget, wealthy countries would need to forego some of their own growth. The need for forbearance by the wealthy figures in many climate treaties, like the Kyoto Protocol. Not surprisingly, this is a controversial proposal and is one reason why the US did not ratify the Protocol.

We can use the disaster game to see if wealthy people will cede development to the poor. In the standard game, the only way for a player to help their group is to contribute to the threshold. But as we've seen, when players differ in wealth, they do not always succeed in preventing disaster (see Tavoni et al. 2011). In our new version, there's another way to help: During development, players can harvest less. When players harvest less, the disaster is easier to avert and less likely to happen. To test whether wealthy players would harvest less, we created a version of the game with another twist. Within a group, we randomly assigned three players to be *old-timers* and three to be *newcomers*. Old-timers, representing wealthy countries, played the development phase as before, getting 10 rounds to harvest. Newcomers, representing developing countries, just watched for the first 5 rounds, and were permitted to harvest only during the final 5 rounds. Old-timers could harvest up to $40 each, while newcomers could harvest a mere $20. As before, every dollar harvested raised the threshold and ate up the carbon budget.

Two brief qualifications are worth mentioning. First, because the groups in this new version could earn less in total than in the previous versions, we made some quantitative changes to how the threshold and risk of disaster worked; the specifics needn't concern us here. Second, the mapping from our game to the real world is not exact. The real-world problem of most interest is the behavior of countries that are *already wealthy*. In the game, the old-timers, who we have called wealthy, do not start out wealthier than other players. Instead, given the design of the game—they have twice as much time to develop—it's very likely that they will end up wealthier. Nonetheless, we think the setup tells us something useful about wealth, development, and restraint, especially in the current context of international climate negotiations.

In principle, old-timers could harvest twice as much as newcomers. But they could also show restraint in their harvests, sacrificing their own wealth to hold down both inequality and the risk of disaster. Did they restrain themselves? As a benchmark, we can compare old-timers' choices to the choices players made in the original version. Both these sets of players could develop over 10 rounds; they only differed in whether they were grouped with other players who did not have this luxury. Let's look separately at the first 5 rounds and the last 5 rounds. In the first 5 rounds, the original players harvested on average 78% of the maximum ($3.12 per round). Over the same span, the old-timers restrained themselves, harvesting only 65% of the max ($2.60 per round). In the final 5 rounds, the original players harvested 79% ($3.14 per round). Over this span, the old-timers again restrained themselves, harvesting 63% ($2.52 per round). Consistent with real policies for allowing catch-up growth, old-timers appeared to recognize their privileged position and reduced their harvests so that some of the carbon budget remained for newcomers. Nonetheless, old-timers did not restrain themselves into equality: Old-timers harvested, on average, a total of $25.60, newcomers only $17.30. (Newcomers on average harvested 86% of the maximum, or $3.46, during each of the 5 rounds they were allowed to develop.)

Although helpful, regulating economic development is just one part of the solution. Keeping the threshold low makes averting disaster easier, but players must still contribute enough to stop disaster. And people might contribute less if they go through different paths of development. Maybe old-timers will not be generous despite their greater wealth, or maybe newcomers will not be very motivated because they face the same problem as old-timers despite not being as responsible for the problem.

To better understand how differing responsibilities affect spending on disaster prevention, we created another placebo version of the experiment. The placebo group was matched to groups of old-timers and newcomers, and each of 12 placebo groups played only the disaster phase. Their threshold, risk of disaster, and individual starting stakes were all drawn from their matched group. So, when the disaster phase started, both the placebo groups and the original groups of old-timers and newcomers were all starting with the same numbers; the only difference was that old-timers and newcomers created their potential disasters while the placebo groups were simply handed theirs.

Did players contribute less when some followed different paths than others? Yes: Whereas placebo groups contributed 52% of their endowments, old-timers and newcomers combined to contribute only 49%. These differences allowed 10 of 12 placebo groups to escape disaster, but only 4 of 12 original groups. Which type of player is responsible for this failure? To meet the threshold, players must contribute 53% of their money. Relative to this, old-timers contributed a bit more, 55%. The problem is that newcomers contributed much less, only 42%. In part, this is because newcomers resented old-timers; through no fault of their own, they had less opportunity to gain wealth and yet were expected to help solve the problem. One newcomer told us, "I decided not to contribute any because I felt that the individuals who were able to [harvest] more money in the first round should contribute more because I started with a disadvantage."

We can summarize how much creating one's own disaster hurts by combining the two games where players created their own disaster and, separately, combining the two placebo games. Doing this, we see that somewhat less than half of self-created groups succeeded (46%, or 11 of 24), whereas nearly all placebo groups did so (92%, or 22 of 24). This translated into tepid earnings for the self-created groups (on average $42, about 40% of the maximum possible), but greater earnings for the placebo groups ($70, about 70% of the max).

Some formulations of common but differentiated responsibilities call on developed countries to take the lead, and we can see if that happens with our data. To do this, we'll take a closer look at what happens with groups that have early and late developers, the old-timers and newcomers. Some of these groups succeeded in averting disaster and some did not (one-third, or 4 out of 12, were successful). What separates the successful from the rest? One possibility is that in successful groups, the old-timers were particularly willing to contribute their fair share or more. And this is what we found: Groups succeeded when old-timers upped their contributions. Recall that we also had placebo groups where players were given old-timers' money *but did not themselves create this money*; the point of the placebo players was that they were just handed the cash. And it turns out that whether or not the placebo "old-timers" contributed extra did not really matter for their group's success. There was something special about real old-timers, the players who created their own greater wealth, leading the way that helped groups prevent disaster. When true early developers took the lead, late developers followed.

Altogether, when it comes to the principle of common but differentiated responsibilities, the evidence from our experiments is a bit mixed. On the one hand, compared to the version where all players are identical and create their own group's disaster, in the version with early and late developers the old-timers restrained themselves from harvesting too much. Old-timers also helped their groups succeed in averting disaster. On the other hand, their groups were not always successful—on average, groups with early and late developers were less likely to stop disaster than placebo groups. Possibly this is because old-timers recognized the basic intuition behind common but differentiated responsibilities yet did not know how to precisely realize it in practice. Exactly how much more should they contribute? What should be expected from newcomers? A similar problem bedevils international negotiations: Countries often cannot decide exactly who should be responsible for what. By allowing countries to make their own decisions, the Paris Agreement side-stepped this problem, paving the way for many countries to sign on. The risk of this, however, is that what countries voluntarily and independently decide on might not be enough. Indeed, research shows that groups with concrete norms are more successful at cooperating (Becker and Ostrom 2003; Cox, Arnold, and Villamayor Tomás 2010).

Self-Created Disaster across Cultures

All four experiments discussed above used student players in our lab at Stony Brook University, New York. Perhaps the behavior we observed was somehow peculiar to our university or to American students more broadly. To check, we replicated our studies at Shanghai Jiao Tong University, China. Specifically, we replicated the two studies that connected the development and disaster phases (so, no placebo replications). As a reminder, in the baseline version all players were identical, with each one having the same number of rounds to harvest from the commons; the other version divided people into early and late developers—old-timers versus newcomers. There were two necessary tweaks. First, we translated the instructions into Mandarin Chinese. Second, the game was played for yuan rather than US dollars. To help readers compare, we report the outcomes in dollars. (To do this, we divide the original yuan amounts by six. At the time of the study, one dollar was worth about six yuan.)

Will Chinese students play like Americans? On the one hand, some research suggests that people differ across countries in how they cooperate and how they play games (Henrich et al. 2005), perhaps because of different national interests or political and economic systems. On the other hand, some research suggests that people have universal, shared foundations of fairness and cooperation, though the exact weight people put on any foundation can vary quantitatively (Haidt 2012). Although we were prepared to find some cultural differences and some similarities, the uniformity we actually found surprised us. Just as in the US, Chinese old-timers harvested much less than both newcomers and players in the baseline game. Thus, old-timers in China also revealed a concern with their special responsibilities. During the first five rounds, baseline players harvested 95% of the maximum ($3.81 per round), but old-timers harvested just 85% ($3.38 per round). In the final five rounds, baseline players harvested 97% ($3.88 per round), but old-timers just 87% ($3.49). Not surprisingly, the newcomers harvested quite a lot during the few rounds available to them, 97% ($3.87).

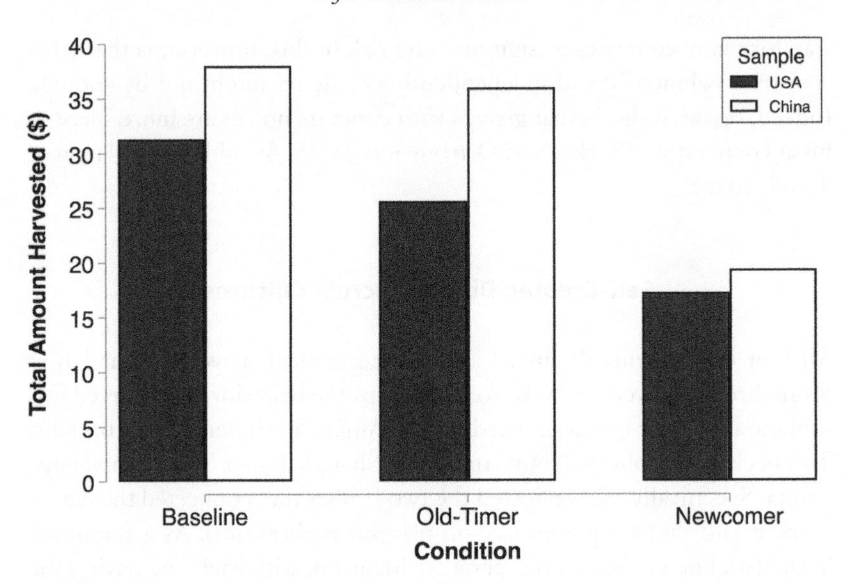

Fig. 22. The average amount each group harvests, by condition and country. Players could harvest at most $40.

Figure 22 summarizes the percentages harvested by country and by experiment. Although Chinese players harvested more overall than Americans, the patterns were otherwise very similar: Early developers showed restraint.

Chinese and American players were also similar when we looked at how much they contributed. In China, baseline players contributed 49% of their money to the threshold; newcomers contributed somewhat less, 40%. The old-timers, however, contributed quite a lot, 55%. (Remember players must contribute 53% to meet the threshold and avert disaster.) Just like in the US, Chinese old-timers were particularly willing to sacrifice for their group, again consistent with common but differentiated responsibilities. Figure 23 summarizes contributions for both countries. Unlike for harvests, for contributions even the overall levels were similar across nations.

Creating Disasters for Others

So far in this chapter, all the experiments have players create disasters that *they themselves* must solve. Climate change, however, is a global dilemma: People routinely take actions that have the potential to create disaster for others. In the previous chapter, we saw several studies that tested whether

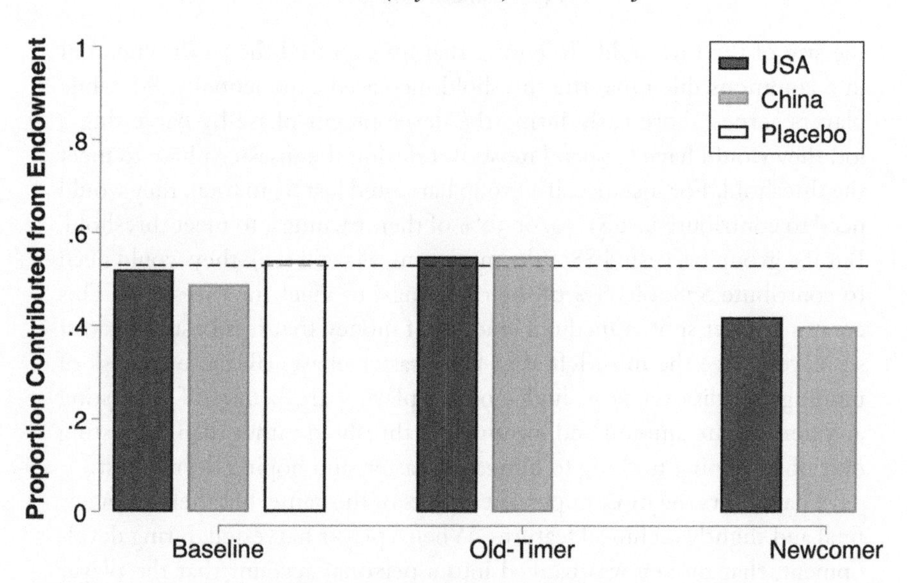

Fig. 23. The average proportion that players contributed from their endowments in each condition. To meet the threshold, players had to contribute 53% of their endowments on average. The dashed line represents this threshold.

people would prevent disasters for others. Here, we examine what people do when their choices create disaster for others. When people do not face the consequences of their emissions, they might be less willing to restrain themselves, even though the emissions still affect others. With our colleagues Alessandro Del Ponte and Nick Seltzer, we created a version of the disaster game that again allowed people to harvest from a commons during a development phase (Del Ponte et al. 2017). The more they harvested, the harder it was to prevent disaster during the disaster phase. The new feature here was that, for some groups, the disaster they created was passed along to a different group.

Passing It along in the Lab

The first version of our pass-along studies was played by groups of 4 students at our university lab. This game was similar to those we described previously: There was a development phase and a disaster phase. Compared to previous studies, we simplified things a bit. The risk of disaster if the threshold was not met was fixed at 90%. Harvesting a lot raised only the size of the threshold. To ensure that we captured the rivalry inherent in a commons dilemma, the threshold increased exponentially. So, while players earned more cash during the development phase by harvesting a lot, they would have to spend most of it during the disaster phase to meet the threshold. For instance, if a group harvested just $5 in total, they would need to contribute just $1.50, or 30% of their earnings, to meet threshold. But if a group harvested $80 (the maximum, as we'll see), they would need to contribute $78, or 98% of their earnings, to meet the threshold. This creates a sweet spot, a medium amount of money that it is best to harvest so players have the most left after the disaster phase. (Because the risk of unmitigated disaster is so high—90%—players are better off harvesting an intermediate amount and meeting the threshold rather than harvesting everything, doing nothing to mitigate disaster, and hoping for the best.)

That covers the most important aspects of the game, but there are a few final and slightly technical features: When a player harvested during development, that money was banked into a personal account that the player could not touch again; although it could be lost to disaster, it could not be spent to prevent disaster. Instead, when the disaster phase started, we gave each player a new pot of money called an endowment. Players could use the endowment to prevent disaster. Any dollar spent from the endowment disappeared regardless of whether the group met the threshold; any dollar not contributed was kept but could be lost to disaster. To determine the endowments for a group, we first calculated the average harvest across all group members. We then gave each player this average amount as their endowment. Here's an example: Imagine a group where players harvested $8, $9, $13, and $14, for a total of $44. These amounts were banked into each player's personal account, never to be touched again. When the disaster phase began, each player received an endowment of $11 (= $44 / 4) that could be used to prevent disaster. This might seem unnecessarily complicated, but why we did it this way will be clear in a moment.

This new game had two conditions. In the self-created condition, when players harvested, they worsened disaster for their *own* group. In the second, pass-along condition, players' harvests created disaster for *another* group. This receiving group, during their disaster phase, received the threshold and the endowments generated by the creating group. (The creating group also had a disaster phase, but the threshold and endowments were created by a third, unrelated group. And the receiving group also had a development phase, which created the disaster phase for yet another, unrelated group.) For the creating group there is no longer a sweet spot in harvests. If they want to maximize their own money, they should harvest everything they can, disaster be damned. The question is whether creating groups, recognizing that they are creating a disaster for others, will show restraint: Will they take everything? Will they show some restraint but not as much as self-created disaster groups? The reason for separate personal accounts and endowments was to ensure that receiving groups could always meet whatever threshold they received. If receiving groups had to use only their personal accounts, it would be possible for them to have too little money to meet the threshold. This would happen if the receiving group showed restraint in their harvests, but their creating group did not and passed a huge disaster along. Using two separate pots avoids this possibility.

In both the self-created and pass-along conditions, groups played the development phase for 10 rounds. Each player could harvest 0, 1, 2, 3, or 4 tokens every round. Tokens were worth 50¢, so players could extract a maximum of $20. After each round, everyone saw how much every other player harvested. Any money harvested was banked in their personal account. Next, players learned the values of their endowments and disaster thresholds. If they were in the self-created condition, this was just the result of their own decisions. Everyone then played the disaster phase for 10 rounds. Each player could contribute 0, 1, 2, 3, or 4 tokens in each round, and they saw how much everyone else contributed after every round. Before the game began, players in both conditions learned the complete structure of the game and how the two phases were connected. In the pass-along condition, we emphasized that groups were *not* trading thresholds.

What did our players do in the development phase? Did those creating their own disaster go for the sweet spot? Did pass-along players show any restraint, or did they take everything and create a major disaster for others? Recall that the most a player can extract is $20. For self-created players, the sweet spot is to harvest most, though not all, of this: $18.60. In fact, self-created players were even more conservative than this, harvesting just $15.31 on average, taking 82% of what they should have. Although they were not spot on, self-created players recognized that it was not in their interest to harvest everything. Pass-along players, however, were not as restrained. They harvested $17.31. This is both good and bad news. On the one hand, they did not harvest everything, as they took only 87% of the selfish maximum. This means they had some concern for their receiving group. On the other hand, using control players' harvests as a benchmark, the pass-along players harvested more than their receiving group would have wished. In other words, pass-along players were restrained, but not completely.

Turning to the disaster phase, what we found surprised us. Essentially every group, regardless of condition, prevented disaster (22 of 24 groups succeeded, or 92%). With almost no one failing, there was no way to use statistics to test whether self-created or pass-along groups had an easier time averting disaster.

Too Much to Lose and Passing It around the World

This finding of most groups succeeding was so surprising that we decided to design another study to follow up. We speculated that a novel feature of our design—splitting money between personal accounts and endowments—was responsible. Canonical versions of the disaster game give players a single pot of cash. From this pot, a substantial portion must be used to avert disaster. In this case, some players might decide it's not worth it to spend on mitigation, and instead take the chance that unmitigated disaster will spare them. In our game, players had two pots and only a part of one was needed to avert disaster. This means that the cost of mitigation is small compared to the total amount exposed to disaster. Given this, players might have felt that they had too much to lose and that they were best off preventing disaster. Our second study was designed to test this idea and another new question, which we'll describe momentarily.

The new study, run over the internet with a sample of American adults, had a few important changes. Internet samples usually play for smaller stakes; in our study, the most that players could earn was $1. Another change was that the development phase and the disaster phase were each just one round. In the development phase, players could harvest any amount, in cents, between $0 and $1; this was again banked into their personal accounts. When players harvested more, they again created exponentially larger thresholds. In the disaster phase, players were given their endowments and from these pots could contribute any amount they wished to their group's threshold.

Two additional changes to the game allowed us to answer new questions. First, we created two different pass-along conditions. In the US pass-along condition, US players created disaster for other Americans (just like our US student sample). In the India pass-along condition, US players created disaster for players in India. Climate change is a global dilemma: People in wealthy countries are primarily the ones who created or are currently creating the real-world analogues of thresholds, yet it is people in developing or poor countries who face the greatest threats. Given this, we wanted to see how relatively rich Americans would treat players from a developing country. On the one hand, Indians are distant foreigners, so Americans might not care to help them. On the other hand, given the clear and direct link between US players' actions and the disaster for Indian players, US players might show similar levels of restraint regardless of whether they create disaster for each other or for people on the other side of the world.

The other change allowed us to test whether having a lot of banked money encouraged players to contribute more to the threshold. We designed three self-created conditions. Each one had a different way of computing endowments for the disaster phase based on how much players banked during the development phase. (Recall that players' banked money was equal to the amount they harvested individually.) In our previous study, endowments were simply equal to the group's average harvests. In our new study, one self-created condition—the equal condition—worked in the same way. Another self-created condition—the doubled condition—endowed each player with twice the average harvest. The final self-created condition—the halved condition—endowed each player with half the average harvest. Thresholds were also adjusted accordingly. (The two pass-along conditions featured only doubling.) If banked money encourages people to contribute to prevent disaster, then people should contribute the largest percentage of their endowments in the halved condition; this condition places a lot of banked money at risk with a relatively small pot used for mitigation. At the same time, people in the doubled condition should contribute the smallest percentage of their endowments; here, most of players' money is tied up in their endowments, so it might seem wasteful to spend a lot of it to protect the smaller personal account. The equal condition should fall in the middle.

To recap, our new study had five conditions. There were two pass-along conditions, one where US players created disaster for Americans and one where US players created disaster for Indians. There were three self-created conditions. They differed in how endowments were computed from average harvests: by halving, doubling, or leaving them the same.

Did players in our new study show restraint in their harvests? In this study, players harvested *very* conservatively. We suspect that this is because they played the development phase in a single round (the original study had 10 rounds with feedback). The sweet spot for harvesting in the self-created conditions ranged from the full $1 in the halved condition, to $0.96 in the equal condition, to $0.80 in the doubled condition. Actual players, regardless of their specific condition, only harvested about $0.50. Players were clearly concerned with harvesting too much and creating too big a disaster.

Turning to the pass-along conditions, we again find evidence of restraint. Players who wanted to maximize their own cash should harvest everything they can. In fact, pass-along players harvested only $0.58, and this was true whether they were creating disaster for other Americans or for Indians. Pass-along players did harvest more than self-created players, about 16% more, but nonetheless they showed considerable self-control. Altogether this replicates our original study: Our new players were worried about making disasters that would be difficult to prevent, even when creating disaster for others.

That our players treated fellow Americans and distant Indians the same is surprising. A lot of research finds that people favor their own groups over other groups. We can only speculate on exactly why our players were so kind. One possibility turns on the fact that all the internet players, American or Indian, came from the same subject pool (called Mechanical Turk, a subpart of Amazon.com). There is some evidence that people in this pool view themselves as part of an overarching group (Almaatouq et al. 2019). Perhaps our American players viewed Indian players as part of their group, rather than distant others. Another possibility is that the game makes abundantly clear how harvesting now creates disaster for others later. This clarity may have inspired players to restrain themselves. If true, this would point to ways that policymakers or activists could generate support for policies of restraint: Show citizens in detail how their choices affect others.

Turning from creating disaster to preventing it, did players contribute enough to stop disaster? Yes: Overall, about 70% of groups met their thresholds. The most important question, though, is how the relative amount of banked money changed contributions. As we expected, players in the halved condition—who had a lot banked relative to their endowments—contributed the most to the threshold, about 31% of their endowment. Players in the doubled condition—with a relatively small amount banked—contributed only 28%. Players in the equal condition were in the middle, contributing 30%. These differences seem small. But remember that small differences can have big effects in threshold games, because thresholds are all or nothing. When players had relatively little money banked (and so contributed little), only about 60% of groups prevented disaster. When they had a lot of money banked (and so contributed a lot), nearly 90% of groups prevented disaster.

Looking at the bigger picture, these findings about banked money are good news for understanding whether wealthy countries will help mitigate disaster. Wealthy countries have lots of "banked money" in the form of infrastructure and other capital investments that can be lost to disasters but cannot easily be spent on prevention or adaptation. For example, you cannot somehow liquidate a skyscraper by breaking it down into its component parts, selling them, and investing that money in wind power. Our findings suggest that wealthy countries could be especially motivated to help because they have so much to lose relative to the costs of mitigation. On the other hand, wealthy countries are overall less exposed to disaster, which might counteract any effect of having more to lose. Figuring out which force is most potent in the real world is, of course, beyond what experiments like ours can test. But our experiments do point one way forward.

An Ethical Lab

Chapter 4 showed that people will prevent disaster for others, and this chapter showed that people try to avoid creating disaster for others in the first place. This was true even though in both cases help comes at a personal cost. Although players were not perfectly altruistic, they were clearly invested in the welfare of others. This illustrates a broader point: Games are laboratory tools for studying not just strategy, but also ethics. Strategy involves making the best choices for yourself, given the choices others are making. Ethics involves, among other things, understanding what the morally right action is. In Chapter 3 on risk and the next chapter, Chapter 6 on trust, the primary issue is strategy, figuring out what's best for players themselves given their knowledge and available choices. In this chapter and Chapter 4, a major theme is ethics: What should players do when their actions can help or hurt others? Ethics are often studied with games because games allow researchers to study questions of fairness or generosity. (Think back to Chapter 2 and the dictator game. The dictator game is a classic because it models a common ethical dilemma—how should someone divide a fixed stake?) Ethics are obviously important for designing good climate policies. Games can help policymakers and citizens understand what people want others to have and when people might pay to ensure that those others get it.

SIX

Trusting Each Other

According to a 2017 Vox headline, "Donald Trump has tweeted climate change skepticism 115 times" (Matthews 2017). In 2013 he tweeted, "We should be focused on magnificently clean and healthy air and not distracted by the expensive hoax that is global warming!" The next year he tweeted, "Any and all weather events are used by the GLOBAL WARMING HOAXSTERS to justify higher taxes to save our planet! They don't believe it $$$$!" And during his presidency Trump declared, "Brutal and Extended Cold Blast could shatter ALL RECORDS—Whatever happened to Global Warming?" Although Trump is an elite, he channels a worry of many citizens: Concerns over climate change are exaggerated or fabricated because it benefits scientists, politicians, and others. Politicians get more tax dollars to control; scientists gain in appointments, prestige, and grant money.

This is an example of *asymmetry*: differences between elites and everyday people in their motivation or knowledge. Elites probably know more about climate change than the public. They might have read the entire IPCC report or had an advisor summarize it for them. And they probably understand the array of policies that could mitigate the worst effects of climate change. Beyond knowledge, elites also have hidden information about their own incentives. Some might care deeply about protecting the environment and stopping climate change; others might be looking for a pretext to create the social programs they already want. It's difficult for citizens to tell the difference, which creates problems for elites who want to enact useful policies. Asymmetry complicates the relationship between elites and the public.

In this chapter we explore three problems that arise from asymmetry. First, politicians can have incentives that push them to enact policies that go against the public interest. Given this, how can citizens figure out which political leaders to trust?

Second, even when political leaders have the public interest at heart, the public often opposes spending on mitigation and adaptation. Other issues, like the economy, seem more pressing. This could in part be because the public lacks information about the seriousness of the problem. How can the elite garner support for useful policies?

Third, just as the public might not trust elites, elites might not trust the public. When politicians enact a policy, they don't know how the public will react. The public could respond in ways that nullify or even reverse a policy's intended effects. We examine the case of geoengineering, a class of technologies designed to radically alter the climate and prevent warming. Some experts believe geoengineering will be a necessary part of mitigation—emphasis on *part*. The public, however, might be overly optimistic about geoengineering, viewing it as a cure-all. Because of this optimism, the public might withdraw their support for piecemeal—but still necessary—mitigation strategies. Or the public might hate geoengineering so much that they'd prefer anything else, no matter how inefficient. When choosing policies to implement, what happens if policymakers do not trust citizens to react appropriately?

In the experiments of this chapter, members of the public play our games. Yet some of the questions are about elites, not the public. As we discussed in Chapter 2, in many ways politicians are just like they rest of us. Elites such as representatives in the US House and members of the Department of State or Department of the Treasury play games similarly to other people (LeVeck et al. 2014). And politicians and citizens use the same heuristics to solve complex problems (Sheffer et al. 2018). Even if elites and citizens differ in some ways, our experiments nonetheless shed light on how everyday people do (and do not!) make mistakes in the face of asymmetry.

Can People Trust the Right Elites?

If you care about climate change, it's easy to spot some untrustworthy leaders. You probably won't listen to Jim Inhofe, the Oklahoma senator who held a snowball on the Senate floor and asked how climate change could be real if it's snowing outside (Bump 2015). And you'll probably ignore Mike

Lee, the Utah senator who claimed the Green New Deal was as fanciful as Ronald Reagan riding a velociraptor; for emphasis, he even brought a photoshopped image of the Gipper astride a Cretaceous refugee (Chiu 2019).

These are fun (and perhaps alarming) anecdotes, but most scientific and political debates about climate change are not so black and white. Usually, people are not arguing about whether the climate is changing. Instead, they argue over problems like how to address climate change economically or how to balance mitigation and environmental conservation. There are some cases where mitigation and conservation are aligned. Preventing rising temperatures not only helps humans avoid swelling seas or increased disasters, it also protects natural habitats and other species. As warming reduces sea ice in the Arctic, polar bears lose their hunting grounds. Within this century, some polar bear populations could disappear (Molnár et al. 2020). And climate change could destroy coral reefs, directly from warming and indirectly by making oceans more acidic (Baker 2001; O'Neill and Oppenheimer 2002). Climate change could also prevent species living in forests from bouncing back quickly after wildfires (Stevens-Rumann et al. 2018). But sometimes mitigation and conservation come into conflict. Consider a proposal to build a new transmission line in Massachusetts. This would allow the state to use more hydroelectric energy (a cleaner source of energy) and to move away from coal power. If the state burned less coal, it would emit less carbon. However, Massachusetts would need to build the line through the White Mountain National Forest, undermining conservation (Roberts 2018a). The governor, an advocate for mitigation, supported the line. But the Sierra Club, a conservation group, opposed it. The line is currently under construction but facing multiple lawsuits as well as staunch public opposition. When it comes to problems like this that have no easy answer, it's difficult for the public to know who to listen to.

Roadblocks to Deciding Who to Trust

Can the public figure out which policies and politicians help, which hurt, and which are useless? At least two major obstacles stand in the way: ignorance and partisanship. First, take ignorance. People seem uninformed of basic facts about politics (Delli Carpini and Keeter 1993). A 2019 poll by CNN found that 12% of Americans had never heard of Mike Pence—yes, Mike Pence, the sitting vice president at the time of the survey (Stieb 2019). If people are this ignorant, how can they be expected to select good politicians or good policies?

Although this problem applies to any political issue, here we're interested in whether the public can navigate the politics of disaster and climate change. Disaster policies must balance investment in *prevention* and *relief*. Prevention includes designing infrastructure to withstand disaster or training first responders. Relief includes providing cash or aid to people in the wake of disaster. Although prevention and relief both help, research suggests that prevention is more effective and efficient (Healy and Malhotra 2009). Indeed, $1 spent on prevention reduces damage from disaster by about $15; in other words, the government would need to spend $15 in relief after disaster strikes to achieve roughly the same result as preemptively spending $1. Plus, prevention preserves things that cannot be recreated no matter how much is spent on relief, things like historic landmarks, fragile ecosystems, and human lives.

While prevention is more effective, it's not as easy to see it in action as relief. In the aftermath of Hurricane Katrina, you probably remember how the Louisiana Superdome sheltered nearly 10,000 people. But you probably never heard how the federal government prepared for storms like Katrina, such as when the National Incident Management System trained 1,200 first responders in Mississippi alone (Congress 2006).

Although prevention is probably more efficient, relief is more visible: Which do voters prefer? Andrew Healy and Neil Malhotra studied whether prevention or relief garners more votes for incumbent presidents in the US. Looking at elections between 1988 and 2004, they found that incumbent presidents received more votes from counties that received more relief (Healy and Malhotra 2009). Prevention spending did not matter. The researchers concluded that voters are shortsighted: Voters simply do not know enough to reward politicians for efficient spending on prevention. Instead, they prefer relief because they can see it in action. In other words, voters are potentially ignorant.

The second obstacle for voters attempting to make good decisions is that they might be blinded by partisanship. Take the United States: Being a Democrat or Republican changes what information voters seek out and believe. People talk more politics with others they agree with (Huckfeldt 2007; Huckfeldt and Mendez 2015), consume news that jibes with their politics (Guess 2021), and give more credence to info that confirms their existing views (Bartels 2002; Campbell et al. 1980). Partisanship also shapes who people trust to tell them about disasters and disaster policy. Democrats tend to follow sources like *NPR* or the *New York Times*, sources that present evidence that climate change is real. Republicans tend to follow network TV news sources that present a more mixed view of climate change (Carmichael, Brulle, and Huxster 2017; Shao and Goidel 2016). Partisans might even change their beliefs to oppose the other party: When Republicans see Democrats on the news urging mitigation support, Republicans are even less likely to believe in climate change (Merkley and Stecula 2021). Democrats are also more likely to believe that climate change has made disasters worse (Boudet et al. 2020). Altogether, research on partisanship and ignorance paints a bleak picture; it's unclear whether citizens can think clearly about complex problems like climate change and related disasters.

Follow the Leader

Because the public knows little, they might cede decisions to leaders (Zaller 1992). In the first game of this chapter, we test this hypothesis. We ask a stark question: Will people follow leaders' decisions even when leaders are no more informed than everyday citizens? In other words, does the mere fact that a person is the leader cause others to follow them?

This study used a variant of the disaster game that allows players to take risks when preventing disaster (see also Chapters 3 and 4). Four players in a group faced an oncoming climate disaster. Each player started with a 20-token personal account (representing liquid capital) that could be invested in mitigation. Each player also had an 80-token personal account (representing illiquid things like infrastructure) that could not be spent on mitigation but would be wiped out by disaster. (Tokens were worth real cash, $0.01 per token.) If the four players collectively contributed 120 tokens to a mitigation threshold, they successfully stopped climate change and kept their remaining tokens. If they failed to meet the threshold, there was a 90% chance that disaster would strike and they would lose their remaining tokens. (Additional information on the methods is in the technical appendix to this chapter.)

Players had three choices for their 20-token account. First, they could defect and keep the 20 tokens for themselves. Second, they could make a certain contribution by adding 20 tokens directly to the threshold, representing investment in piecemeal but certain mitigation technologies like solar power. Finally, they could make a risky contribution, representing investment in technology like geoengineering. If players made a risky contribution, there was a 50% chance their 20 tokens were doubled to 40 before adding to the threshold and a 50% chance their contribution would fail, contributing nothing. Either way, the player forfeited their personal account when they contributed it.

Players were randomly assigned to one of two conditions. In the *control condition*, they played the game as described above. Each player decided simultaneously and with no communication what to do with their 20-token account. In the *leader condition*, one player was randomly chosen as the leader. The leader made their contribution first and their choice was broadcast to the other three players, the followers. With no further communication or information, the followers decided whether and how they wanted to contribute.

How should players contribute in the control condition? Because the threshold totals 120 tokens, the group maximizes its earnings if all four players make risky contributions. If three out of four gambles pay off, the group contributes enough to stop disaster (3 * 40 = 120). Having all four players gamble maximizes the chance that at least three will succeed. There are other defensible courses of action (including everyone defecting), but none as good as four risky contributions (for the game theory, see Andrews, Delton, and Kline 2018). In fact, the plurality of control players, 48%, did make a risky contribution. Close behind, 43% made certain contributions. And a scant 9% defected (see Figure 24).

How should leaders play? It turns out that leaders should contribute just like players in the control condition: Everyone earns the most if the leader makes a risky contribution. Consistent with this, we found that actual leaders behaved just like control players: As shown on the left in Figure 24, leaders' choices matched control players' choices. For instance, 48% of control players made risky contributions, as did a nearly identical 51% of leaders.

How should the followers respond to leaders? We'll take in turn the three possibilities: the leader makes a risky contribution, makes a certain contribution, or defects. If the leader makes a risky contribution, this does not change the game: Followers should also make risky contributions. In this instance, following the leader is the best way to avert disaster.

There's no good reason for the leader to make a certain contribution, but if they do, how should followers respond? In the best course of action, one follower makes a certain contribution and the other two make risky contributions. If both risky contributions succeed, the four contributions meet the threshold of 120 (= 20 + 20 + 40 + 40). Here's the upshot: Compared to followers who see a leader make a risky contribution, followers who see a leader make a certain contribution should themselves be more likely to make a certain contribution.

Finally, if a leader defects, what should followers do? Although other choices could be justified, by far the most obvious is for everyone else

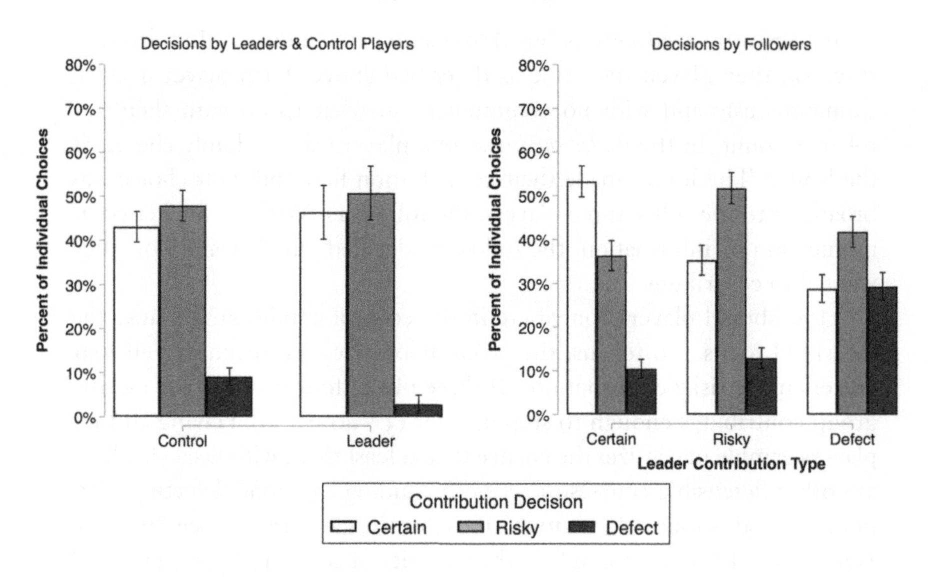

Fig. 24. The proportion making certain, risky, and no contributions in each condition. Error bars are standard errors of the mean.

to defect as well. Remember, even if everyone defects there's still a 10% chance that everyone keeps their money.

All in all, players should follower their leader, and they do: As shown on the right in Figure 24, when the leader makes either type of contribution, followers match it, following risky with risky and certain with certain. When the leader defects, followers are more likely to defect than after any other choice by the leader. Unlike contributions, however, there's never a majority of followers defecting, even if their leader defects. Followers prefer cooperating even when the leader does not.

In this game, leaders were selected randomly and thus had no special knowledge about the game, but the other players followed them anyway. Why? One possibility is that the followers treated leaders' choices as focal points (Schelling 1960). Focal points are things, places, or ideas that make intuitive sense for everyone to coordinate on. For instance, if you and a friend agreed to meet in New York City but forget to specify where and cannot contact each other, the most obvious place to meet might be Times Square. If you forgot to specify a time, noon is most obvious. When the group needs to come to a set of decisions—should we defect or contribute?—the leader helps reduce the possibilities by selecting one.

The leaders in this game, however, had no opportunity to bluff about their incentives or intentions. Whatever choices leaders made were faithfully broadcast to the followers. But things are not always so clear; leaders can exaggerate or lie. When dissembling is possible, will people still follow the leader?

Institutions to the Rescue

It's difficult for people to tell which politicians are trustworthy, but not impossible. One way to tell is if the politician is constrained by a good *institution*. Broadly, an institution is a set of rules that a leader (or others) must follow. Some institutions have formal rules (like a legal code), some informal rules (like social norms), and many have both. When the public knows a leader is constrained by a good institution, they can trust the leader, regardless of whether the leader is dishonest at heart, because the institution prevents the leader from being dishonest (Muller and Mestelman 1998; Ledyard and Szakaly-Moore 1994; Noll 1982; Lupia and McCubbins 1998; Boudreau 2009). Not all institutions are good; some create a *stake in inefficiency*. Leaders often organize publics goods. An institution that creates a stake in inefficiency gives the leader the opportunity to exaggerate the cost of a public good and profit from excess contributions to that good (Miller and Hammond 1994).

We can see this distinction in two carbon taxes put forward by Washington state. The first, Initiative 732, did not generate a stake in inefficiency. The tax on carbon emissions was balanced by a cut to the sales and business and occupation taxes. As a consequence, the tax did not generate any new revenue, so politicians could not use any new revenue for political gain. The second proposed tax, Initiative 1631, created the potential for a stake in inefficiency. The initiative levied a carbon tax and proposed investing that money broadly in environmental conservation and green energy. But this initiative did not commit these funds to specific conservation or energy plans, so legislators had discretion over how to spend them (Roberts 2018b). Under this policy, politicians could use the money generated by the tax to fund projects that benefitted themselves. Imagine you're deciding how to vote on these two initiatives. You probably don't know all the details, but you might know enough to see that Initiative 1631 gives politicians far more wiggle room than Initiative 732. For that reason, you might prefer the latter. Indeed, Washingtonians rejected Initiative 1631 in part on the grounds that it gave too much discretion to politicians in how to use the tax revenue (Bailey 2018; Bernton 2018a, 2018b).

Now, think back to the finding that voters reward politicians for spending on relief, but not on prevention. Instead of showing that voters are short-sighted or uninformed, another interpretation is that voters are concerned about stakes in inefficiency (Gailmard and Patty 2018). It's difficult to know, in advance, whether prevention money is being spent wisely. Maybe politicians are spending on things that will reduce the damage of disaster, or maybe they're shoveling money to their cronies. Is building this new seawall the most efficient way to keep back the waves, or is the construction company owned by a friend of the mayor? When politicians spend on relief, however, it's clearer where the money is going. Paying to rebuild houses wiped out in a flood is a straightforward case of helping disaster victims. (Rebuilding work could, of course, be handed to cronies, but it's still work that needs to be done and therefore still a clearer use of money than prevention.) In our next experiments we test whether people notice the difference between institutions that do and do not constrain politicians. Are people less willing to spend on prevention when political leaders can exploit such spending?

Manipulating Institutions

We again used a disaster game with a leader. The twist here is that we manipulated whether the leader had a stake in inefficiency. In the inefficiency condition, the politician could profit from excess contributions to the public good.

We ran two versions of this experiment, one with a sample of American adults who played over the internet through Amazon's Mechanical Turk (MTurk), and one with students who played in our lab. Besides the setting, the only difference between the two experiments was the amount of money at stake: The online players could earn a maximum of $1.35 and the students in the lab $27.

Each group had four *citizen* players. Together they faced an oncoming flood that would wipe out all their money. The flood was imaginary, but it would really destroy their money. Their money was denominated in experimental tokens, and each player began with 135 tokens (for online players 1 token = 1¢; for students 1 token = 20¢). Each citizen simultaneously decided how many tokens to contribute to build a levee and thereby prevent the flood. If they contributed enough to build the levee, they kept all their remaining tokens. If they didn't build the levee, they lost everything.

Unlike in typical versions of the disaster game, here the citizens did *not* know exactly how much it cost to prevent disaster. They were only told that the cost was randomly drawn from 80, 160, 240, 320, and 400 tokens. But one player did know the exact cost: We added a fifth player, the *leader*. The leader could not contribute to build the levee but would lose their tokens if disaster struck. (The leader started with only 75 tokens. We gave them this smaller starting stake because they could not contribute; this keeps their final payout roughly the same as the citizens'.) After we told leaders the real cost of the levee, they were allowed to send a computerized message to tell the citizens what they learned. What they said was entirely up to them. They could tell the citizens the true cost of building the levee, or any of the other possible costs. After reading the leader's message, citizens each simultaneously and independently decided how much to contribute. This is like disaster prevention in the real world. For example, those living on the southeast coast of the US might know that infrastructure like levees will help prevent damage from hurricanes. However, everyday citizens probably don't know exactly how much it costs to make levees effective. It is up to engineers to identify these costs, and then politicians to build the support needed to fund the levees.

Also like outside the lab, the leader in this game could lie. When and why would the leader choose to misrepresent the costs of disaster prevention? Our interest here is the ability of institutions to control or prevent dissembling by politicians. So, we manipulated the institution that determined the leaders' final payoff such that sometimes leaders had a stake in lying, sometimes not. In the *control condition*, the leader kept their 75 tokens if the citizens contributed enough to build the levee and, importantly, they had no opportunity to earn any additional tokens. In the control condition, the leader should always tell the truth. It was in their best interest to help the citizens coordinate on the correct cost to stop the flood.

In the *inefficiency condition*, the leader similarly had 75 tokens they stood to lose if the citizens didn't build the levee. But they could also skim some money from the citizens' contributions to disaster prevention. If the citizens contributed more than the cost of the levee, the leader took one-fourth of these extra tokens for themself. For example, if the citizens contributed 100 tokens, but the levee only cost 80 tokens to build, the leader received their initial 75 tokens and 5 extra tokens, one-fourth of the 20-token excess. According to the game theory, in the inefficiency condition, the leader should always lie and say it costs 400 tokens to build the levee.

After the leader told citizens the (potentially misleading) cost of building the levee, citizens did two things. First, they told us what they believed to be the true cost of the levee. If they were correct, they earned extra money, though players only found out whether they had been correct after the entire game was over. If they trusted the leader, they should say the true cost is equal to what the leader said; if not, they should say the true cost is less. Next, the citizens, simultaneously and with no communication with each other, contributed toward the levee. If contributions met the true threshold, disaster was averted; if not, everyone was *guaranteed* to lose everything. (This was different from most disaster games because usually there is some chance of avoiding disaster even without meeting the threshold.)

Our key question is: Are citizens sensitive to the stake in inefficiency? That is, can they distinguish between different institutions to decide whether to trust the leader and support disaster prevention? Political scientists are divided on the question of whether, in the real world, voters are equipped to do this. Voters are ignorant of many political facts, they may be biased by their partisanship, and they may hold incoherent beliefs about politics (Campbell et al. 1980; Converse 1964; Kuklinski and Quirk 2000; Delli Carpini and Keeter 1993). All of these would keep them from having a critical perspective on political institutions. Yet other researchers think that voters do approach politics in a broadly rational way (Miller and Hammond 1994; Lupia and McCubbins 1998). Although voters may not understand all the nuances of a political issue or institution, they know enough to get what they want—and most voters do not want corruption. Indeed, some anthropologists and psychologists argue that humans have evolved to delegate decisions to leaders and that, co-evolved with this, humans scan for cues that their leaders are misbehaving (Garfield, von Rueden, and Hagen 2019; Case and Maner 2015; Boehm 2009; Price and Van Vugt 2014).

Making Predictions in the Game

Before turning to the results, we need to unpack a bit more about the game. First, what should players do if they totally disregarded the leader's message? Based on how we set up the game, if citizens believe the leader's message is unreliable, they should act as if the threshold was the middle of the five possible thresholds (i.e., 240 tokens; see Andrews, Delton, and Kline 2022b). Then, they should contribute enough to meet this threshold. We assume that the most attractive way to do this is for players to divvy up the costs equally as in previous versions of the disaster game, each contributing one-quarter of the threshold. To make this math easier for the

TABLE 9. Each possible cost of the threshold, as well as citizens' fair-share contribution for each threshold cost. The final two columns show whether citizens should believe the leader if they send each possible levee cost in each condition.

Threshold Cost (In Tokens)	Fair-Share Contribution (In Tokens)	Should citizens Trust the Leader at This Message?	
		Control Condition	Inefficiency Condition
80	20	Yes	Yes
160	40	Yes	Yes
240	60	Yes	Yes
360	80	Yes	**No**
400	100	Yes	**No**

participants, we told them exactly how much their fair-share contribution would be at each possible threshold (Table 9).

Second, what should players do in the control condition? In this condition, the interests of the citizens and the leader are completely aligned. This means leaders have no incentive to lie, and citizens should trust whatever the leader says. So, citizens should contribute to meet whatever threshold the leader says.

Finally, what should players do in the inefficiency condition? Now the interests of citizens and leader are not aligned. The leader is best off telling citizens that the cost is as high as possible (i.e., 400 tokens), and the citizens should disregard this claim. Thus, citizens should contribute as if the threshold were right in the middle of the possible range.

Things are bit complicated in the inefficiency condition, though, because what citizens should do depends on the leaders' actual message. If the leader says that the threshold is one of the highest two costs, the logic we just gave holds: Citizens should ignore the message. But if the leader says the cost is in the middle of the range or below, citizens *should* believe the leader. Consider what it means if the leader says it only costs 80 tokens to build the levee—the leader doesn't benefit from saying the cost is so low. If the leader is lying, saying the threshold is lower than it actually is, the citizens might not contribute enough, and everyone, including the leader, will lose everything. Additionally, the leader doesn't get to benefit from the contributions above the cost of the threshold. So, if citizens see a low message, they should assume it came from a leader who is ignoring their (the leader's) own best interests and acting on behalf of the group. In the inefficiency condition, citizens should treat low messages as honest. Table 9 summarizes when citizens should trust the leader. In sum, they should trust all leaders except those who are in the inefficiency condition *and* who claim the threshold is very high.

What Did Players Do?

Recall that we had adults play this game over the internet and college students play in our lab. Despite the different settings, both sets of people played the game similarly. One thing stands out immediately: Regardless of whether players were in the control or inefficiency condition, most of them trusted their leader. Out of all players, 41% said that they believed that the leader told their group the true threshold. Because of this trust, when leaders sent messages saying the threshold was higher, citizens contributed more.

Nonetheless, did players take the institution into account? Yes. Although players in both conditions trusted leaders less when they said the threshold was high, this was especially pronounced in the inefficiency condition (see the left panel of Figure 25). As examples, let's look at what happened when leaders said the threshold was at the maximum (i.e., 400 tokens). Here, 32% of citizens in the control condition trusted leaders, but in the inefficiency condition only 20% did so. In other words, when the institution did not constrain the leader, trust dropped by nearly 12 points as leaders in the inefficiency condition sent messages saying the levee is more and more expensive.

This mistrust then undermined contributions to disaster prevention. In the right panel of Figure 25, we show the proportion of citizens who contributed their fair share. At the top two thresholds, there were differences among citizens: Compared to the control condition, players in the inefficiency condition were less likely to contribute their fair share. For instance, when the leader said it cost 400 tokens to build the levee, 61% in the control condition contributed enough to meet this high cost; in the inefficiency condition, only 51% did so.

Politicians send lots of conflicting messages about how to address climate change and prevent related disasters. Some bring snowballs to the Senate floor as proof the earth is not warming, while others propose carbon taxes. We find optimistic evidence that people can recognize and appropriately respond to institutional incentives that make it more likely or less likely for someone to misrepresent the costs of disaster prevention. People can identify which leaders to trust in the face of disaster. People might therefore support prevention spending when they are more confident that those funds will be used only to help prevent disaster.

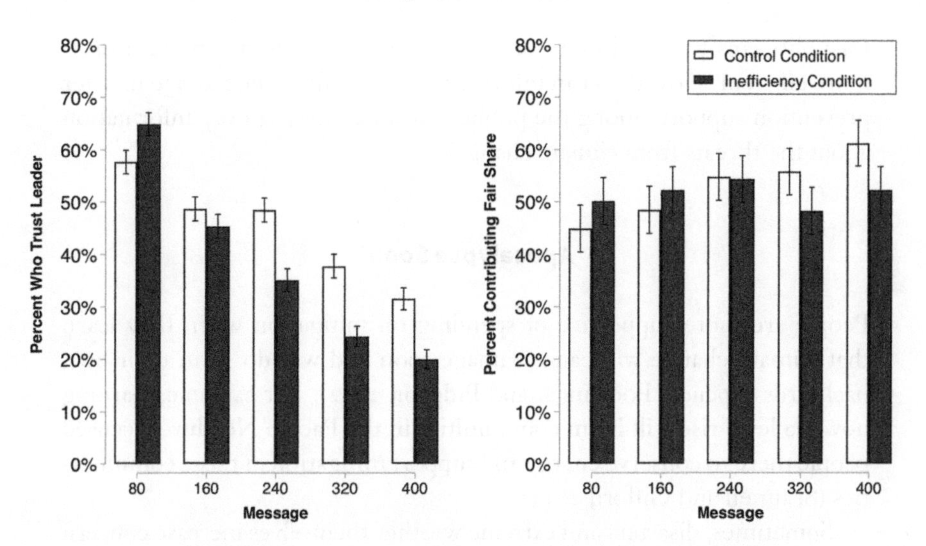

Fig. 25. The first panel shows the proportion of players who believed the levee costs exactly what the leader says. The second panel shows the proportion who contributed their fair share to meet the threshold. Error bars are standard errors of the mean.

Of course, outside of inefficiency incentives, there are other reasons political leaders might misrepresent the costs of prevention. For example, they might strategically try to send more funds to different districts to secure votes for upcoming elections (Sainz-Santamaria and Anderson 2013; Reeves 2011). And citizens might be ideologically opposed to some forms of disaster prevention (Friedman 2019). Or they might believe that some disasters are simply not important enough to be worth preventing. We turn to this problem next.

Encouraging More Prevention Spending

Designing good institutions will probably encourage support for mitigation, but alone it is not enough. Even when political leaders have the public's interest at heart, the public undervalues spending on disaster prevention (Motta and Rohrman 2019; Healy and Malhotra 2009), and a majority of Americans do not prioritize spending on mitigation or adaptation for climate change (Egan and Mullin 2017). These preferences are consequential in democracies because public opinion shapes policy. For example, presidents who spend more on disaster relief are more likely to be reelected than those focused on prevention. This creates an incentive for politicians to spend more on relief, even when disaster prevention is more efficient. How then can informed elected officials encourage disaster prevention support among the public? How can they convey information about the threats from climate change?

Apocalypse Soon!

People are more supportive of spending on mitigation when they learn that climate change will cause damage soon and will do so in their own backyards (Spence, Poortinga, and Pidgeon 2012). For example, learning how sea level rise will harm communities in the Pacific Northwest caused people there to conserve energy and support mitigation in those communities (Scannell and Gifford 2013).

Sometimes, disasters and extreme weather themselves increase concern about climate change. Imagine that you are a skeptic. You aren't sure the earth is warming and, even if it is, you doubt that rising temperatures will affect you and your family. But then your home is hit by Hurricane Sandy. Sandy was the second-costliest hurricane to make landfall in the United States and was responsible for the deaths of 285 people. Not only was this hurricane severe, it was unique in how far it traveled up the coast (Blake et al. 2012). After you witness its destruction, you learn that Sandy was made worse by climate change (Cody et al. 2017). What happens to your opinions about climate change?

Or perhaps you lived through the polar vortex in the northeastern United States in 2017. Or the severe California wildfires in 2018. Or the 2013 flooding in Colorado. It doesn't matter exactly which disaster you lived through; firsthand experience will probably help convince you that the earth is warming and is causing problems now. Indeed, research finds that people who experience extreme weather, or even just changing weather, often become more concerned about climate change and more willing to spending on mitigation (Shepard et al. 2018; Rudman, McLean, and Bunzl 2013; Ray et al. 2017; for review, see Howe et al. 2019).

There are a few reasons that experience changes minds. First, people usually imagine climate change as a *distant* threat, likely to harm only people on the other side of the globe. Thus, our own government should not prioritize it (Leiserowitz 2005). Seeing the damage for oneself makes climate change seem less distant (Spence, Poortinga, and Pidgeon 2012; Weber 2013). Second, disasters engage people's emotions. Imagine watching a hurricane leave a train of destruction as it travels up the coast—this makes people anxious about future disasters. When people are scared, they are more willing to support prevention spending (Albertson and Gadarian 2015; Atkeson and Maestas 2012; Slovic et al. 2004; Finucane et al. 2000).

A final problem is that people may believe that if the damages *caused* by disasters are expensive, then disaster *prevention* is also expensive. That is, they might engage in what we call *cost conflation*: believing disasters with expensive consequences would have been expensive to stop. Why might they have such beliefs? When people are missing pieces of information, they look for patterns and draw connections between potentially unrelated factors (Nickerson 2002; Bar-Hillel and Wagenaar 1991). This is especially true when people are trying to avoid losses (Pligt 1998). People don't know exactly how much it costs to stop climate change, but they do learn the cost of disasters as we face them more and more. If people engage in cost conflation, they will assume the two are related. To test whether cost conflation explains why people support mitigation after being exposed to the costs of climate change, we turn to a study using the disaster game that was conducted with our colleague John Barry Ryan (Andrews and Ryan 2021).

Cost Conflation in the Disaster Game

To test whether people believe expensive problems have expensive solutions, we created another version of the disaster game. In this experiment, university students played in our lab. In groups of four, they faced an oncoming disaster, represented as usual by a monetary threshold. Each player had 40 tokens (1 token = 10¢). If they failed to prevent the disaster, they paid a penalty out of those tokens; this represented the damages caused by the disaster. However, if they met the threshold, they kept their remaining tokens.

The game was played over a computer network, and the computer randomly decided for each group the penalty for failure. Players were told that the computer would choose a penalty of between 1 and 30 tokens. Once the computer chose, it revealed the exact penalty to players. Thus, everyone knew whether the consequences of disaster were cheap, expensive, or somewhere in between.

The twist here is that players did *not* know the size of threshold to prevent the disaster. We told them that the threshold was too large to be met by a single player, yet small enough that a group working together could meet it. And—as we carefully explained to them—there was no connection whatsoever between the penalty for failure and the size of the threshold. The computer determined both these numbers independently of each other.

The game began with the computer randomly and independently choosing both the penalty and the threshold. The computer then communicated the penalty to the players. Each player next guessed the size of the threshold. If they guessed correctly, they won an extra 5 tokens (however, they were only told whether they were correct once the experiment was over). After this, the four players simultaneously and independently decided how much to contribute to the threshold. Finally, they learned whether they met the threshold, and paid the penalty if they failed.

This setup mimics key features of climate change mitigation. The IPCC argues we must keep the rise in global temperature below 2°C to avoid the worst of climate change (IPCC 2014), but it's difficult even for experts to know what sacrifices are needed to prevent this rise (Roe and Baker 2007). So, we don't know exactly how expensive the threshold is to prevent disaster. The penalty cost, then, represents the information the public receives about how bad climate change will be if we fail to stop it.

Did our players fall victim to cost conflation, believing greater penalties meant the threshold was larger? Yes: As shown in the left of Figure 26, when they faced a larger penalty, players believed that the threshold was larger. This was true *even though we told them the two values were unrelated.* Despite explicit information to the contrary, players assumed expensive problems had expensive solutions. And these perceptions changed players' behavior: As shown on the right of Figure 26, when the penalty was bigger, players also contributed more. Specifically, this panel shows whether players contribute more or less than what they should contribute to maximize their earnings; as the penalty increases, players contribute more than the payoff-maximizing amount.

The Downsides to Cost Conflation

You might be thinking, great! We have an obvious way to encourage support for mitigation—just make the costs of unmitigated climate change seem even worse! This seems to be the strategy taken by an article published in *New York Magazine* in 2017. The article, titled "The Uninhabitable Earth," opens by saying, "Famine, economic collapse, a sun that cooks us: What climate change could wreak—sooner than you think" (Wallace-Wells 2017). Popular media often highlight the damage from disasters from climate change (Feldman, Hart, and Milosevic 2017). Is this the right strategy to get people motivated? Probably not.

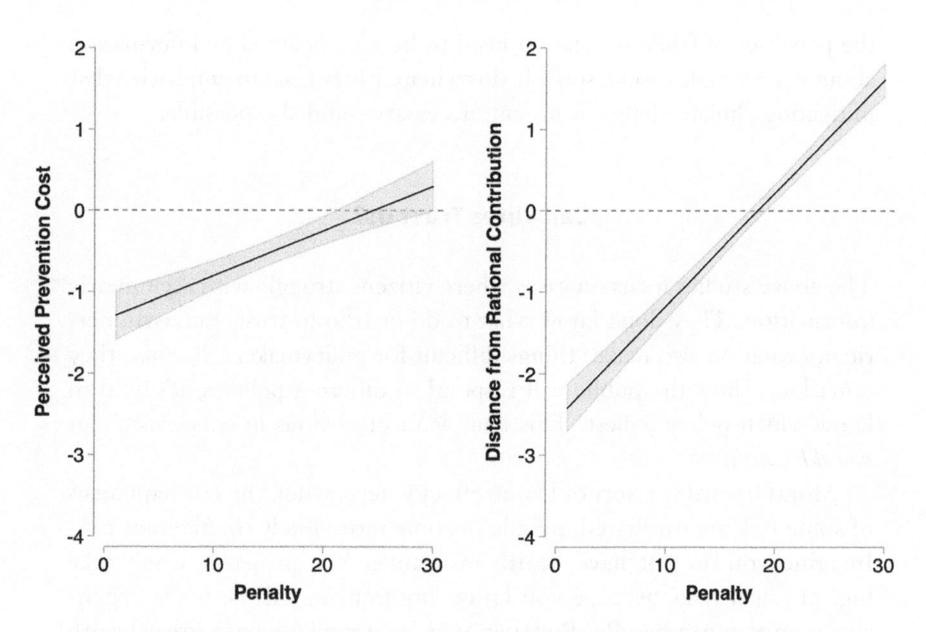

Fig. 26. The first panel shows the distance between how much someone guessed the threshold cost, and what a payoff-maximizing player should guess the threshold cost. The second panel shows the distance between how much someone contributed to meet the threshold, and what a payoff-maximizing player should contribute to the threshold. The x-axis shows the size of the penalty. As the penalty increased, people believed disaster prevention was more expensive, and they contributed more to disaster prevention.

In the disaster game on cost conflation, we found two reasons for concern. First, cost conflation led our players to contribute more than necessary. In the real world, this kind of inefficiency would leave other important needs unfunded. And, as we discussed above, inefficiency opens the door for corruption (Gailmard and Patty 2018). Second, if we emphasize just how terrible unmitigated disasters could be, people might think they are too expensive to solve. Under cost conflation, if unmitigated climate change will be apocalyptic, the costs of prevention must be astronomical. Beliefs like this undermine mitigation support. Research using surveys finds that emphasizing disasters associated with climate change can overwhelm people to the point that they no longer want to think about the issue (Feinberg and Willer 2011; Andrews and Smirnov 2020; Levine and Kline 2017).

How then can elites generate support for mitigation? As we discussed in Chapter 3, the consequences of climate change need to be severe enough to mobilize costly mitigation (Milinski et al. 2008). But messages about the penalties of failed mitigation need to be accompanied by information about what people can reasonably do to help. Elites need to emphasize that mitigating climate change is urgent, necessary—and also possible.

Can Elites Trust Us?

The above studies focus on cases where citizens struggle with asymmetric information: They don't know what to do or who to trust. But asymmetric information also makes things difficult for policymakers. Because they don't know how the public will respond to different policies, it's hard to know which policy is best. This issue is most obvious in concerns about *moral hazards*.

Moral hazard is a sort of ironic effect where, when the consequences of some risk are mitigated, people become more likely to take that risk. Imagine you do not have health insurance. You probably won't take big, physical risks because you know any injuries will be more expensive than you can handle. But then you get a new job with great health insurance. Now you climb mountains, skydive, and go running with the bulls—you take wild risks, because you know insurance will cover any injuries. You've fallen into moral hazard *if* the new risks you take outpace the new backstop provided by insurance. In other words, it's a moral hazard if you become so risk-loving that your probable injuries go beyond what your insurance will cover. (Different people use moral hazard to mean slightly different things; see Jebari et al. 2021. We use it to cover situations where people become overly optimistic about how much their risk is mitigated.)

In the context of climate change, many pundits worry that elites will create moral hazard among citizens if elites use, or perhaps even mention, *geoengineering*. As discussed in many places throughout this book, geoengineering covers a lot of strategies. Generally, it refers to intentional, large-scale attempts to change the climate. Adding a few more wind farms would not be geoengineering. Seeding the atmosphere with aerosols to reflect light back to space would be geoengineering. Improving the efficiency of batteries for electric cars would not be geoengineering. Capturing huge amounts of carbon from the air and storing it would be geoengineering. Critically, geoengineering includes negative emissions technologies (i.e., capturing carbon from the air). In fact, any path described by the IPCC for avoiding catastrophic change in the climate requires these negative emissions technologies (IPCC 2021). Methods for carbon removal include things as technologically simple as afforestation (that is, planting trees) or as complex as creating machines that suck carbon from the air. These methods all demand difficult tradeoffs. For example, afforestation competes with agriculture for land. And, as discussed in Chapter 4, they could backfire and cause more harm than good.

However, when it comes to the moral hazards of geoengineering, the problem is not the technology itself but how people might respond. There is a lot of uncertainty about how successful geoengineering will be, but even optimal geoengineering alone is not enough to stop climate change. Yet, when describing geoengineering, popular media use headlines like "The climate moonshot: engineering the earth" (Okutsu, Maulia, and Phoonphongphiphat 2021) or "Carbon capture could save the planet" (Brown 2019). If citizens think that geoengineering efforts will totally fix the climate problem, they have little incentive to support other mitigation efforts like investment in solar and wind power, despite the fact that these investments are still necessary. So, while geoengineering would help us avoid damage from climate change, it might simultaneously undermine support for other required efforts.

Many elites, from policymakers to experts on mitigation and public opinion, are worried that creating or even discussing geoengineering could cause moral hazard in citizens (Amundson and Biardeau 2018; Barrett 2007; Bodansky 1996; Hale 2012; Kolbert 2014; Lin 2013; Reynolds 2015; Scott 2012). Yet, when researchers have studied how real people respond to the technology, they mostly (but not always) find evidence that moral hazard is not a problem. After reading about geoengineering, people still support investment in other types of mitigation (Corner and Pidgeon 2014; Fairbrother 2016; Kahan et al. 2015; Merk et al. 2016; Pidgeon et al. 2012). One experiment finds that telling people about geoengineering actually causes them to want *more* investment in other mitigation strategies like wind and solar power (Merk, Pönitzsch, and Rehdanz 2016). Survey respondents hated the idea of geoengineering so much that they were willing to pay more to stop policymakers from using it!

Despite scant evidence, if policymakers expect people to engage in moral hazard, then policymakers may withhold funding for geoengineering—even when it's safe and effective. We call this *moral hazard anticipation*. This is a problem of asymmetric information: Because elites don't know how the public will respond, they might guess incorrectly and withhold mitigation technology that would help.

Identifying Moral Hazard and Moral Hazard Anticipation

We looked for both moral hazard and moral hazard anticipation in a modified disaster game (Andrews, Delton, and Kline 2022a). To our knowledge, no researchers have tested whether people engage in moral hazard anticipation, despite the risk that needless worry about moral hazard will undermine investment in geoengineering. Additionally, research looking for moral hazard in response to geoengineering has typically used survey experiments. In these studies, the researchers briefly describe geoengineering. What researchers choose to include or leave out in these descriptions could have big effects on how people respond, especially since people generally don't know much about geoengineering beforehand (Mahajan, Tingley, and Wagner 2019). By using an economic game, we can mimic the strategic problems of geoengineering and moral hazard that people might face in the real world. Games have drawbacks, of course, but they do provide a new way of looking at the question. Given what previous research has found and the way elites talk about geoengineering and moral hazard, we expected to find that people do *not* engage in moral hazard itself but *do* engage in moral hazard anticipation.

In this version of the game, we again have four players taking the role of citizens who must decide whether and how much to contribute to mitigation. A fifth player takes the role of policymaker. The policymaker cannot contribute directly to threshold. Instead, the policymaker decides whether to implement geoengineering; using it costs them and the group nothing. In this experiment, *geoengineering cannot backfire*. (See Chapter 4 for a version that does allow backfiring.) If geoengineering is successful, it completely solves the problem of climate change. If it fails, nothing happens—the citizens face the same problem they did before. In principle, elites should always use geoengineering: In this game it has only upsides and no downsides. However, if policymakers worry that the citizens are overly optimistic, they might withhold geoengineering. Admittedly, real geoengineering can backfire and cannot solve the whole climate problem. We made these stark design choices to make it maximally likely that we would see moral hazard (Raimi et al. 2019): Most evidence suggests that people won't engage in moral hazard (and we do not expect them to either), so we wanted to give moral hazard its best shot. This approach also allowed us to isolate the effect of any moral hazard anticipation.

We recruited participants online through MTurk and assigned them to groups of five, with four citizens and one policymaker. The four citizens each had an endowment of 100 tokens (1 token = 1¢). They faced an oncoming disaster that would with certainty wipe out their tokens. But they could collectively stop this disaster if they contributed 160 tokens to the threshold, 40 tokens on average per citizen.

The policymaker began with an endowment of 60 tokens, which would also be lost if disaster was not averted. Unlike the citizens, the policymaker could not contribute to the threshold. Their primary task was to decide whether to use geoengineering. First, the policymaker learned the exact probability that geoengineering would succeed. This probability was randomly drawn by the computer and could be 10%, 30%, 50%, 70%, or 90%. Second, the policymaker decided whether to use geoengineering.

The citizens then learned *only* whether geoengineering was used. They did *not* learn the probability that it would succeed, nor did they learn whether it actually succeeded. (They did know the possible probabilities of success.) The citizens then simultaneously and independently decided how much to contribute.

If citizens engage in moral hazard, they will contribute *less to the threshold when the policymaker uses geoengineering*. That is, even though in this game they are always best off contributing (because they don't know how likely geoengineering is to succeed), if they are *overly* optimistic about geoengineering's success they will not contribute to stop disaster. However, if citizens are appropriately calibrated to the problem they face, they should always contribute enough on their own—regardless of what the policymaker does. (This is not universally true but was a choice on our part. It depends on the exact probabilities and amounts at stake; for the math, see Andrews, Delton, and Kline 2022a).

If policymakers engage in moral hazard anticipation, *they will be less likely to use geoengineering, but only when it has a low probability of success*. Here's the logic. First, consider what happens when citizens are victims of moral hazard. This means they will contribute too little when the policymaker uses geoengineering. This should not matter to the policymaker when geoengineering is likely to work; when geoengineering's success is virtually certain, it doesn't matter what the citizens do. But at low chances of success, the policymaker needs the citizens to contribute; geoengineering probably won't get the job done even if it is used. Second, consider what happens if citizens are not victims of moral hazard. Here, the policymaker should always use geoengineering. It can't hurt anything, but can only increase the chance that the group averts disaster, so there's no downside to using it, just in case.

So, do citizen players engage in moral hazard? Nope: The left panel of Figure 27 shows that it did not matter whether the policymaker used geo-

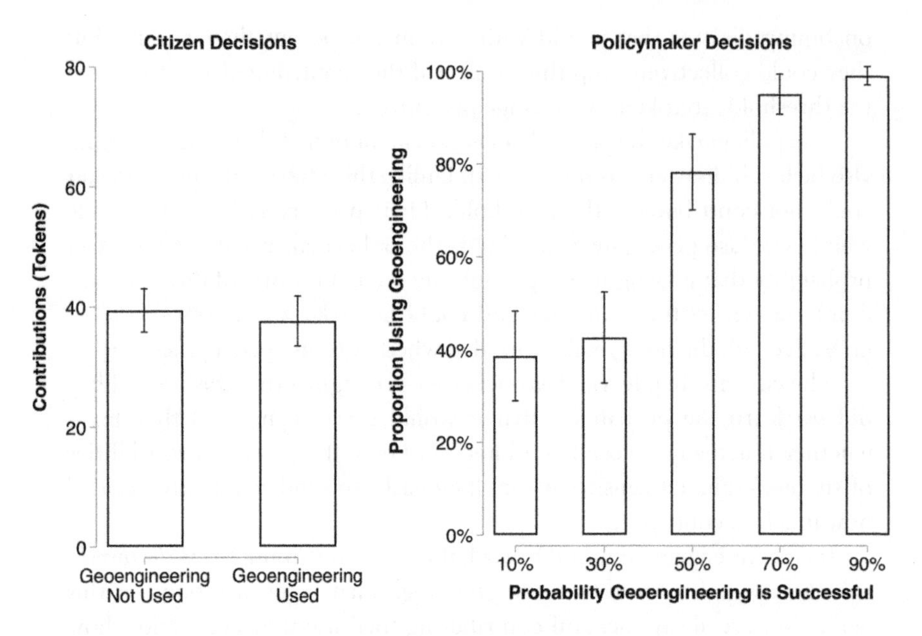

Fig. 27. The left panel shows the proportion of policymakers who decided to use geoengineering, for each probability that geoengineering succeeds. The right panel shows the average contributions made by citizens toward the threshold when geoengineering was and was not used. Error bars are standard errors of the mean.

engineering. Regardless, the citizens contributed the same amount. This is consistent with the bulk of the surveys testing for moral hazard. Moral hazard does not seem to be a problem.

Even though there is no moral hazard, do policymakers *anticipate* that there will be? Yes: We found policymakers did not trust their citizens to behave appropriately. As shown in the right of Figure 27, they withheld geoengineering when it had a low probability of success, even though geoengineering could only help the group avert disaster. At the highest chance of success, a whopping 99% of policymakers used geoengineering; at the lowest chance, only 39% used it.

Did Citizens Know Too Much?

Two features of our game might make it hard for us to find moral hazard. First, although citizens did not know the exact probability that geoengineering would succeed, they knew it was 10%, 30%, 50%, 70%, or 90%. In principle, this allows them to reason that, even when geoengineering is deployed, they are still best off contributing to help prevent climate change. Outside of the lab, however, people know very little about geoengineering (Pidgeon et al. 2012; Corner, Pidgeon, and Parkhill 2012). Maybe because of this ignorance, outside of the lab people will engage in moral hazard. Second, in the study we just described we mentioned geoengineering in the game instructions; it was our research question, after all. But perhaps just mentioning geoengineering made people too optimistic, regardless of whether the policymaker used it. Or maybe people fear geoengineering and any mention of it makes them worry that the problem is unsolvable.

To address both these issues, we designed a second experiment with a few changes for the citizens. First, citizens were randomly assigned to know or not know the possible probabilities that geoengineering would succeed. All citizens were told whether the policymaker used geoengineering, but only half knew that the probability it would succeed was randomly drawn at 10%, 30%, 50%, 70%, or 90%. Those who didn't have this information could therefore make wild guesses about the probability of success and might be more optimistic. Thus, those with less information could be more likely to engage in moral hazard. Second, we introduced a control condition where we didn't mention geoengineering. In this condition there was no policymaker at all—four citizens simply faced a threshold of 160 tokens, and each had to contribute enough to stop disaster. This allows us to see whether just mentioning geoengineering undermines contributions.

Do we see moral hazard now? No: We again find no evidence that citizens engage in moral hazard. As illustrated in Figure 28, there were no meaningful differences in how much citizens contributed regardless of whether geoengineering was used and regardless of whether they knew the possible probabilities that it would succeed. And it only barely mattered whether we mentioned geoengineering: Citizens generally contributed 50 tokens in every condition, but they contributed just a bit less in the geoengineering conditions compared to the control condition. The difference is very small, though, so we suspect it doesn't mean much.

As discussed in Chapter 4, there are several reasons to be concerned about geoengineering. Our results here, however, suggest that moral hazard should not be among these concerns. These results also show that it's not just citizens who suffer from asymmetric information: True, citizens can struggle to trust leaders, but leaders can fail to trust citizens.

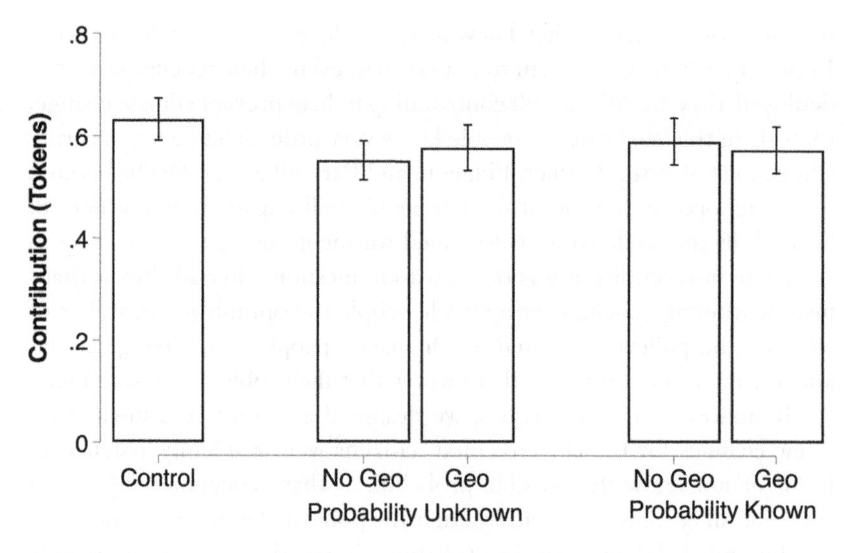

Fig. 28. Average contributions to the mitigation threshold across experimental conditions. Error bars are standard errors of the mean. "No Geo" = No geoengineering used; "Geo" = Geoengineering used.

Trust in Disaster

People receive conflicting signals about climate change, including about the damage that disasters are likely to inflict. In 2019, when Hurricane Dorian drew closer to the US coast, the National Weather Service said it would not make landfall in Alabama. Yet for some reason, President Trump tweeted the opposite and even included a map that had been doctored to show Alabama in the hurricane's path (Smith 2019). While this is an extreme example, even people arguing in good faith disagree about the kinds of disasters that climate change will produce.

Politicians and citizens who are eager to help must nonetheless grapple with information asymmetry. Leaders often have access to more information about climate change, but they can also have incentives to exaggerate or lie. And the public has their own private information—leaders cannot perfectly predict what their constituents will do.

Using the disaster game, we find optimistic evidence that people can use institutional cues to identify which leaders to trust. If a politician can personally profit from prevention spending, citizens are more likely to oppose the policy. Policymakers need to design policies that constrain leaders and give citizens confidence that they are not being exploited.

Our cost conflation experiment shows that when people learn that climate disaster will be costly, they are more likely to spend on mitigation. This is not a free pass, though, for politicians or journalists to spread tales of apocalypse. Citizens must feel that the problem can actually be solved. This requires a delicate touch in science and policy communication, not a sledgehammer of negativity.

Finally, despite the concerns of many pundits, we find optimistic evidence that citizens are not victims of moral hazard when it comes to geoengineering. The problem of asymmetric information, and the purely hypothetical problem of moral hazard, should not delay the implementation of critical technologies for mitigation, including safe and effective geoengineering.

Technical Appendix

This appendix provides more details on the follow-the-leader experiment reported in the main chapter.

Additional Methods

Players were 479 people who played the game online through Amazon's Mechanical Turk (MTurk). Our sample consisted of only US adults who had not previously participated in any of our experiments using the disaster game. They received $0.50 just for participating and started with a $1.00 pot of bonus money that they stood to lose if they did not collectively contribute enough to prevent disaster. The game was played just once (i.e., it was a one-shot). Players made their decisions independently and without communication; indeed, players played asynchronously. So, although we matched players into groups for payoffs, all players were statistically independent for analysis.

In all conditions, players in groups of four faced oncoming disaster. Each player had a personal account of 20 tokens and an endowment of 80 tokens. If they collectively contributed 120 tokens to their threshold, they prevented the oncoming disaster and kept their remaining tokens. If they failed to contribute enough to meet the threshold, there was a 90% chance they each would lose their remaining funds. As a reminder, players could only contribute their 20-token personal accounts. They could do so either as a certain contribution, which brought the group 20 tokens closer to meeting the threshold, or as a risky contribution, which had a 50% chance of bringing the group 40 tokens closer to meeting the threshold and a 50% chance of adding nothing to the threshold. They could also defect and simply keep the 20 tokens.

Participants were randomly assigned to one of two conditions. In the first, the control condition, players in groups of four simultaneously decided whether and how to contribute to their group's disaster threshold. In the second condition, the leader condition, one player was randomly selected to be the leader. The leader made their contribution while knowing it would be broadcast to the other three players in the group, the followers. For the followers, we used the "strategy method" (see Chapter 4): They reported how they would contribute for each possible choice of the leader. That is, they separately told us what they would do if the leader made a risky contribution, a certain contribution, or no contribution to the threshold. They were then matched into groups, and payoffs were calculated using the actual decisions of the leaders.

Results

Do leaders make different decisions than those in the control condition? To test this, we conducted *t*-tests to determine whether there are differences in the rates of certain, risky, and no contributions between leaders and those in the control (see Figure 24 in the main chapter). The results, presented below, illustrate that leaders and those in the control condition contribute similarly. They were both most likely to make a risky contribution, followed by a certain contribution, and only a handful defected.

Do people follow their leader? Figure 24 in the main text suggests they did. In the following table we test this with a series of *t*-tests, identifying

TABLE 10: *T*-tests on differences in contribution behavior between leaders and those in the control condition. The first column shows the different contribution types. The second shows the proportion of each contribution type in the control condition, while the third shows the proportion among leaders.

	Condition			
Contribution Type	Control	Leader	*t*-statistic	*p*
Certain	0.48	0.51	0.41	0.68
Risky	0.43	0.46	0.48	0.63
Defect	0.09	0.03	1.69	0.09

TABLE 11. ***T*-tests identifying differences in contribution behavior in response to leader contributions. The first column shows the different contributions followers could make. The second and third columns show the proportion of followers who make each contribution when the leader does or does not make that same contribution, respectively.**

Follower Contribution	In Response to a Leader Contribution of . . .		*t*-statistic	*p*
Certain	Certain 0.53	Not Certain 0.32	5.11	< 0.001
Risky	Risky 0.52	Not Risky 0.39	2.98	0.003
Defect	Defect 0.29	Not Defect 0.12	5.50	< 0.001

whether followers were more likely to make each contribution decision when the leader made that same decision.

In each instance, followers were swayed by the leader—they were more likely to make a certain, risky, and no contribution when the leader did. If the leaders made a certain or risky contribution, the most common response by followers was to match the leader. Defection was different, however. Although followers were more likely to defect when the leader defected, it was never the most popular response; only a third of followers defected even when their leader did.

SEVEN

Looking Beyond the Lab

Climate change is an especially wicked problem (Incropera 2016). For example, it's a disaster of our own making. Those most at risk are often those without the resources to mitigate the risk. And running through everything is a great deal of uncertainty. There is uncertainty about the costs of mitigation, about the potential damages, and about the choices our friends, fellow citizens, and leaders will make in response.

But one thing is certain: Climate change is one of the most pressing issues of our time. Our decisions today determine which future we face. The Intergovernmental Panel on Climate Change discusses these branching possibilities in terms of Representative Concentration Pathways (RPCs); each pathway assumes different amounts of greenhouse gases are emitted over the next century. One pathway is "business as usual" (called RPC8.5). In this pathway, no emissions reductions occur, causing temperatures to rise an estimated more than 4°C, with accompanying rises to sea levels and serious threats to agriculture. The most extreme reduction pathway looks very different (RPC2.6). It involves rapid, intense decarbonization now to avoid later climate disaster. And there are pathways between these two extremes, each one trading off willingness to mitigate and adapt now against damages in the future. Which path we follow will depend on how quickly we cut emissions and invest in technologies such as carbon dioxide removal, technologies that could buy us time to transition to less carbon-intensive energy.

These choices are complicated by the strategic problems of climate change we have seen throughout this book. Climate change is a global social dilemma—we're all best off if we cooperate, but each of us also has an incentive to free ride off the contributions of others. With uncertainty, first in identifying the most effective strategy to stop climate change, and second about whether others will help share the costs, it may not surprise you to hear that some are drawing inspiration from unlikely sources. For example, eruptions like that of Mt. Tambora, which temporarily cooled the planet (see Chapter 1), have inspired intentional efforts to seed the atmosphere with chemicals. Such actions might temporarily slow global warming, but they could also have serious unintended consequences (Andrews et al. 2023). On the other hand, because climate change is a social dilemma, there are concerns that people won't be willing to invest in *any* large-scale attempt to fight climate change. Motivated by concern over lackluster responses to climate change, in 2017 David Wallace-Wells published "The Uninhabitable Earth" in *New York Magazine*. In this widely read—and controversial—article, he argues that "no matter how well-informed you are, you are surely not alarmed enough," citing mass famines, plagues, wars, and deaths that will probably result from our warming planet.

Despite the many roadblocks, our experimental results consistently leave room for optimism. People want to cooperate, they find effective solutions, and they can identify leaders to trust when navigating complex climate policy. These optimistic findings are reflected in recent climate policy. For example, the Inflation Reduction Act promises billions of dollars to fund the expansion of renewable, green energy in the United States. As Wallace-Wells himself said in a follow-up to his pessimistic climate article, "We're getting a clearer picture of the climate future—and it's not as bad as it once looked" (Wallace-Wells 2019). The future, though still uncertain, is not as bad as it once looked *because* of people's social, political, and economic responses. Understanding the factors that influence these responses is the chief concern of this book.

Finding Solutions in the Lab

Even as we make choices today about addressing climate change, we will face new choices in the future. Perhaps we delay mitigation now, forcing future people to invest in riskier technology. Or perhaps we invest now in sustainable energy like wind power only to create future problems due to competition over the use of land for either energy generation or agriculture. Games provide insights into how people might respond to current climate threats and policies, but they also can help us understand how people might respond to dilemmas in the future.

As we discussed in Chapter 2, games require researchers to (1) identify a real-world situation they want to study, (2) come up with a laboratory environment that captures the key strategic elements of that situation, and (3) develop a theory to explain how people should or actually do behave. Importantly, games are not limited to problems that we currently face. Take, for example, the question of how to govern solar radiation management, various technologies that slow the effects of climate change by reducing the amount of sunlight that warms the earth. This technology does not exist right now, at least in a way that it could be used to effectively stop climate change. And there are passionate, ongoing debates about whether it should be further developed, and even whether it should be researched at all. Many of these concerns are technological: For example, will solar radiation management disrupt precipitation patterns? Could it have dangerous interactions with the atmosphere? These concerns are about unintended physical consequences. Some worries, however, center on human behavior; these can be studied using games. For example, a common concern is that if we deploy this technology, or even just invest in research on it, then many citizens will take it as a cue to stop investing in other types of costly mitigation strategies. In other words, people might be overly optimistic about the promise of solar radiation management, a perspective sometimes known as a belief in "technosalvation." According to critics, even if solar radiation management goes off without a hitch, it could still make us worse off because governments, businesses, and individuals would stop supporting policies such as transitions to more carbon-free energy.

We can test some of these concerns in games. When the potential for technosalvation is on the horizon, do people abandon other costly—but necessary—strategies to prevent disaster? We find they do not (Chapter 6). Are people over-enthusiastic about easy fixes, even when they could backfire or even when decision-makers are insulated from any consequences of their choices? Not particularly, we found (Chapter 4). In this case, games can't tell us whether solar radiation management will disrupt weather systems or fall short of promised temperature reductions. But they can suggest how people could respond to the technology, even before it exists.

Of course, games have their limits, and at least two concerns should be kept in mind. First, how does behavior in the game translate outside of the lab? A strength of games is that we can strip down the complexity of the real world into something tractable—something where we can see exactly what is going on and know everything about the constraints and incentives facing players. For example, we might wonder about relationships between leaders and citizens in disasters. If political leaders have considerable flexibility in how they can spend money on disaster prevention, will constituents trust them less? This question is difficult to answer with observational data because so many other things are happening in the real world that might make constituents mistrust their leaders. Are constituents responding to spending flexibility or, say, changes in the economy or other aspects of local policy? If observational research finds that people oppose prevention spending, is that because they are shortsighted or because they don't trust their leaders to implement prevention effectively? Both concerns might lead to the same outcome—voting a politician out of office. How can we unravel this? Games allow us to isolate possibilities and test them one by one. If we think the ability of leaders to benefit will reduce citizens' trust, in a game we can manipulate just the possibility of corruption and see what happens. If we think citizens will pull back from mitigation when geoengineering is available, we can manipulate just geoengineering's availability.

This is great, but a downside is that it leaves important questions unresolved. Outside of the lab, competing influences may sometimes pull in the same way as our mechanism of interest, and at other times they may push against it. So, even if we know in that lab that people trust less when leaders have too much flexibility, it's hard to know exactly what the effect will be outside the lab–other considerations of course affect support for disaster prevention. For example, do you believe you are vulnerable to disaster? If a hurricane is barreling toward you, you might support even the most inefficient spending if you thought it would protect you. How does the nature of the risk matter? Some research finds that people want more government intervention when the risk is seen as unfair to the victims (e.g., child abuse, terrorism) compared to partially being the victims' fault (e.g., health risks from smoking and alcohol abuse; see Friedman 2019).

The second concern to keep in mind is whether the game effectively captures what it's meant to. Here, we have focused on the disaster game, which captures the social dilemma of climate change. While we are all better off if we can avert climate change, we all have an incentive to free ride off the mitigation efforts of others (Keohane and Victor 2016). The disaster game also captures the threshold nature of climate change—beyond an (uncertain) threshold of temperature rise, there will be irreversible damages (Lenton 2011). The different permutations of the disaster game are designed to examine different pieces of the puzzle in the context of social dilemmas and disaster thresholds. For example, does inequality make cooperation more difficult? What happens when there is uncertainty? Can people make good decisions for others?

But this is not the only way to describe the strategic problem of climate change. Instead of viewing it as a social dilemma, some argue we should focus on the *distributive conflict* of climate change (Aklin et al. 2020). This emphasizes that climate change and its resulting policies create winners and losers, and that actual choices about climate change can be explained by conflict over who wins and who loses. In this view, effective policy will only be enacted if those who stand to gain from mitigation have more political power than those who will lose. This approach downplays worries about cooperation: Nations enact climate policy when it benefits themselves, regardless of whether they see others doing so as well.

We agree with statistician George Box, who famously said, "All models are wrong but some models are useful" (Box 1979). No game captures all the complexities of climate change. In fact, if they did, they wouldn't be useful! A game that includes everything would be just as difficult to understand as the real world. Instead, we believe each game presented here models a crucial aspect of the climate puzzle. Games are models that are necessarily wrong; they do not capture all the interpersonal, environmental, or political forces that shape behavior surrounding climate change. Nonetheless, they are useful because they allow us to observe how people respond to specific and identifiable changes in the incentives and institutions they face, and to the choices others make. A variety of models help shed light on the climate problem, including models of collective action and models of redistribution. And a variety of games together provide insights into potential climate solutions.

Of course, games are just one tool for understanding how everyday people and policymakers behave in the face of climate change. In games, players take the imaginary worlds we create as a given, accepting the game of strategy as it is regardless of their real-world beliefs. For instance, some of our players leave us comments at the end of the games revealing that they are extreme climate change deniers;[1] nonetheless, they play the games the same as everyone else. This is why researchers also need survey experiments, observational data, and other sources of evidence. Survey experiments present people with facts and messaging about actual (not simulated) climate change, to see how it affects their beliefs and their desires for policy. These kinds of experiments point the way toward policies that citizens would generally support. And there are useful methods beyond experiments. Qualitative interviews and case studies go in-depth with small sets of people to assess how they are thinking about, or responding to, climate change. Case studies can also uncover the policy challenges, and potential opportunities, for successful adaptation and mitigation. Games are useful but cannot provide a complete picture. Other methods are also necessary.

So, what have we learned with our games? We close by summarizing five lessons, many of which surprised even us.

1. People Cooperate.

In nearly every experiment presented in this book, people have an incentive to not cooperate and instead to free ride off the contributions of their group. Sometimes they can earn a bit more money in the end by paying a cost up front to cooperate—but the less risky move is still defection. In other games, it is clearly more expensive to cooperate, and each person would be better off not helping, even if refraining hurts others. Yet *in not a single study do we see more than a tiny handful of people defect entirely*. People are willing to contribute when the threshold is so expensive it would take the whole group's cooperation to stop disaster. People are willing to contribute when they don't have much to lose, or don't have a high chance of losing. People are willing to contribute when they personally face no risk at all, but their contribution could help someone else stop disaster.

We are not the first to document the astounding willingness of people to help each other, even at a personal cost. Indeed, entire research programs in psychology, economics, political science, biology, and more are devoted to the mystery of what makes us so willing to help one another. Although many animals cooperate and help each other to some degree, humans have evolved to be cooperators par excellence (Raihani 2021). But we have also evolved to live within relatively small-scale societies—on the order of tens, hundreds, or perhaps a few thousand other people. And we have evolved only to think about relatively small timescales—our own lives, our children, and our grandchildren. Our minds did not evolve to think about how billions of people, acting over centuries, could accidentally reshape the earth's atmosphere and climate (Gifford 2011). We hope, nonetheless, that humanity's evolved abilities for cooperation, combined with enlightened self-interest, will allow us to overcome these global problems. Research revealing the overwhelming willingness of people to cooperate suggests this may be possible.

As we tour other insights from this book, keep these high rates of cooperation in mind. When we talk about features that undermine cooperation, it is never to the point of no cooperation.

2. People Want Their Cooperation to Be Effective.

The second (and also optimistic) takeaway is that people not only want to cooperate, but they want to do so effectively. What exactly does this mean? Some of our more cynical colleagues have asked if people are cooperating in games to feel good about themselves. Maybe that feeling is worth more than the few dollars they could save by defecting. Indeed, previous research has argued that people are willing to pay a cost to help others because it gives them a "warm glow" (Andreoni 1995). Are people only helping because it makes them feel good about themselves, or perhaps because it makes them look good in front of others?

We do not deny that self-interest is one reason people cooperate, but self-interested cooperation isn't necessarily a bad thing. We need people to reduce their carbon emissions and support climate change mitigation and adaptation. It only becomes an issue if the behaviors that make people feel good about themselves lead to behaviors that are ineffective or even harmful. For example, there is some evidence that people prefer to support climate change in very public, visible ways, and are less likely to help protect the environment in private (Brick, Sherman, and Kim 2017). This suggests that many people do not care much about climate change but want to feel good about themselves or make others applaud them for their largesse.

In our studies, though, we find that people are willing to spend both money and mental energy to help in the way that is most useful. They want to be effective. For example, even when people personally don't like taking risks, they are willing to do so when it is their group's best chance to stop disaster (Chapter 3). People are also willing not only to spend money to help others prevent disaster, but to spend extra money to make sure they help in the best way possible (Chapter 4). People aren't throwing money away in these games so they can say they helped; they actually want to protect themselves and others from disaster.

Games aren't the only method that suggests people want to cooperate effectively. Among those who care about climate change and want to help, a major barrier to action is not knowing the best way to do so (Axelrod and Lehman 1993; Doherty and Webler 2016; Chu and Yang 2020). If you give these people information about how to help, they often follow that advice (DeSombre 2018; Ajzen 1991). While giving people more information isn't the best strategy to turn the apathetic into climate warriors (Albertson and Busby 2015), information can make a big difference among those who already want to act.

Together, these first two takeaways should inspire some optimism in our fight against climate change. People (generally) want to do good things and seek out information to make sure they're doing so in an effective way. Of course, other considerations get in the way during our lives outside the lab. You might care about climate change, but when deciding whether to drive to work or bike you might instead be focused on how bad the weather is. When deciding which candidate to vote for, you might take into account their climate policy, but you might also be worried about their stance on gun control, education policy, or other economic issues. To translate people's willingness to cooperate into action, they need clear, reliable information on what to do.

3. Clear Coordination Points Can Mean the Difference between Success and Failure.

People are much more willing to cooperate when others cooperate too (Fischbacher, Gächter, and Fehr 2001). And this applies to climate change: People support mitigation policies in their own country when they see that other countries are also investing in mitigation (Tingley and Tomz 2014). But remember that climate change is a special problem because there are thresholds: If we emit too much carbon dioxide or let the planet get too warm, disaster is highly likely. So, we need people and nations not only to cooperate with each other, but to cooperate *enough* that they avoid these disaster thresholds.

Across experiments, we find that clear coordination points—mutually recognizable indications of how much everyone must contribute—make it much easier for groups to avoid disaster. Surprisingly, uncertainty isn't always a big problem, at least for some types of uncertainty. People seem happy to contribute even when they cannot tell exactly how bad disaster would be if it occurs. On the other hand, if people are uncertain about how much is needed to stop disaster in the first place—that's catastrophic (Chapter 3).

Why this difference between uncertain effects of disaster versus uncertain costs to prevent disaster? In every chapter, we found that people generally contributed their fair share or close to it. (Generally, a fair share just means an equal amount of the total cost.) If everyone knows what their fair share is, and everyone knows that everyone else knows, it's easy to cooperate effectively if you want to (Deutchman et al. 2021). But when people don't know the threshold, how could they possibly know what their fair share is? In their ignorance, they might try to guess the threshold (Chapter 6) or simply not contribute (Chapter 3). Either way, people in ignorance of the threshold are much less likely to meet it and avoid disaster.

These results highlight the critical role of climate science and science communication outside of the lab for identifying and broadcasting these coordination points. We've discussed in detail the need to avoid too great a rise in global temperatures, but there is ongoing work trying to define "too much" (Lenton 2011). Organizations like the IPCC are doing critical work to identify this information and make it available to the public, enabling global coordination. However, coordination points also exist at the local level, for example identifying the best strategies for adapting to local impacts and the costs of those strategies. There are ongoing arguments about what role science should play in mitigation and adaptation, in part because stopping climate change is as much a political as a technological problem (Sobel 2021; Drake and Henderson 2022). The results from experiments with climate games illustrate the opportunity for scientific results to coordinate our efforts to overcome climate problems.

4. Inequality Makes Cooperation Harder, but Clearly Defined Responsibilities Help.

We are not all equally at risk from climate change, and we do not all have the same tools available to fight it. This is true at the individual level: Low-income residents of New Orleans are more likely to live in low-lying areas, exposing them to more risk from hurricanes (Colten 2006). This is also true at the international level: The United States has more resources to adapt to the changing climate than island nations like Kiribati that are already sinking under rising sea levels. What are the consequences of this inequality for the prevention of climate impacts?

Inequality isn't always a problem: If people have access to unequal resources to fight climate change, they still generally can meet the threshold, partly because rich players contribute quite a lot (Chapter 4 and 5). When there is inequality in the risks people face, we find they similarly band together to avert disaster. However, when there is inequality such that the richest players are also the *least* at risk, cooperation falls apart (Chapter 4). Unfortunately, this situation is closest to the world outside of the lab.

Rich players are often unwilling to help poorer, more at-risk players. How can we reconcile this with our findings that people are willing to pay to help completely unrelated groups or to restrain their emissions to help future generations? We think the difference is the presence or absence of clearly defined responsibilities. In our own experiments where players can pay to help another group, they are the other group's only chance at success. The second group cannot help themselves and cannot be helped by anyone else. When players give away their own money to reduce the risk of disaster for future groups, again they are the future groups' only shot. When everyone has the same resources and vulnerability, an equal, fair-share contribution is a clear and intuitive focal point. However, when there is inequality, it isn't immediately clear exactly how much each player should contribute. Some might argue that the most at-risk players should pay a bit more than an equal share, since they benefit the most. Others might argue that rich players should pay more, since they have more to give. Different moral intuitions point in opposite ways. Without clearly defined responsibilities, in the end everyone contributes too little, and the group fails to prevent disaster.

Unlike with coordination points, attributing responsibility is not just a matter of science but a matter of ethics and values. According to the United Nations Framework Convention on Climate Change, nations should have common but differentiated responsibilities, where those who have benefitted the most from historic emissions are the most responsible for paying the costs of emissions reductions. But, in part because of this, one of the biggest emitters (the United States) has been reluctant to join international mitigation agreements. Attributing responsibility to bear the burden of mitigation is not easy, but these games reveal how critical agreeing on an answer will be for successful mitigation.

5. People Are Sensitive to the Incentives They Face.

Everyday people are more likely to be maligned than applauded for their political sophistication. Research often finds that people know little about politics and have trouble seeing past their own partisan affiliation (Campbell et al. 1980; Delli Carpini and Keeter 1993). As an example, research finds that people oppose carbon tax rebates because they don't know how much money they stand to gain and because their political party opposes the rebate (Mildenberger et al. 2022).

We find optimistic evidence that everyday people are better at navigating politics than these studies might suggest: *They are sensitive to the incentives they face*. What exactly does this mean? Across experiments, despite different political affiliations, varying climate change attitudes, differences in education, and more, we find that people are good at navigating complex games and are able to identify effective strategies to avert disaster.

When deciding which strategies to use to meet the threshold, players were surprisingly good at figuring out when to take risks (Chapter 3). Despite concerns that people will be overly optimistic in the face of a possible technosalvation like geoengineering, we find no evidence that people are unwilling to support other expensive mitigation strategies (Chapter 6). When taking the role of policymaker, they were willing to deploy simulated geoengineering if it was the group's best chance of success (even if they then wrote us angry messages about how geoengineering outside of the lab is a bad idea). And people were able to figure out which leaders to trust and who to be more skeptical of based on how leaders were compensated. They rightfully distrusted those who had an incentive to exaggerate the cost of stopping disaster (Chapter 6).

This doesn't contradict findings that people are often confused or overly partisan about politics. But it suggests people aren't as blind to the realities of climate change as some worry—and there is ample evidence of this outside of the lab. When people experience climate disasters that highlight the dangers of climate change, or even notice changes in the weather, they are more concerned about the issue (Howe et al. 2019). They reelect politicians who build wind turbines in their districts, rewarding them for both the local economic benefits and the broader benefits of renewable energy for the environment (Bayulgen et al. 2021; Urpelainen and Zhang 2022). Of course it's concerning when oil companies try to push blame for climate change onto individuals, for example by urging them to calculate their own carbon footprints (Leber 2021). But, when responding to their own incentives and the incentives of major polluters, our work suggests that people can see through the ruse.

• • •

These experiments on disaster prevention leave us optimistic, but not Pollyannaish. People are both generous and broadly rational in disaster games. But this is because they buy in to the games, accepting the rules and incentives as they are. This causes even self-proclaimed climate change sceptics to engage in mitigation. In our view, creating a shared understanding of the real world, about the climate and beyond, is the thornier problem. Creating such an understanding will require insights from the academy and beyond.

Note

1. For example, respondents have left us comments about their belief in climate change such as "I just wanted to let the researchers know, I admire the audacity of the people still pushing this farce and outright lie," and "There are three kinds of people that believe in Climate Change: 1) The Socialist Ideologue that understands the REAL underlying agenda; 2) The 'Low Information' person that accepts the false narrative, which is full of half-truths and manipulated data, because they are too intellectually 'lazy' to practice true critical analysis; 3) The DUMBMASSES."

Bibliography

Agee, Ernest, Jennifer Larson, Samuel Childs, and Alexandra Marmo. 2016. "Spatial Redistribution of U.S. Tornado Activity between 1954 and 2013." *Journal of Applied Meteorology and Climatology* 55 (8): 1681–97. https://doi.org/10.1175/JAMC-D-15-0342.1

Ajzen, Icek. 1991. "The Theory of Planned Behavior." *Organizational Behavior and Human Decision Processes* 50 (2): 179–211. https://doi.org/10.1016/0749-5978(91)90020-T

Aklin, Michaël, Matto Mildenberger, Sarah Anderson, Patrick Bayer, Mark Buntaine, Iza Ding, Jessica Green, et al. 2020. "Prisoners of the Wrong Dilemma: Why Distributive Conflict, Not Collective Action, Characterizes the Politics of Climate Change." *Global Environmental Politics* 20 (4): 4–27. https://doi.org/10.1162/GLEP_A_00578

Albertson, Bethany, and Joshua William Busby. 2015. "Hearts or Minds? Identifying Persuasive Messages on Climate Change." *Research & Politics* 2 (1). https://doi.org/10.1177/2053168015577712

Albertson, Bethany, and Shana K. Gadarian. 2015. *Anxious Politics: Democratic Citizenship in a Threatening World.* Cambridge University Press.

Almaatouq, Abdullah, Peter Krafft, Yarrow Dunham, David G. Rand, and Alex Pentland. 2019. "Turkers of the World Unite: Multilevel In-Group Bias among Crowdworkers on Amazon Mechanical Turk:" https://doi.org/10.1177/1948550619837002

Amir, Ofra, David G. Rand, and Ya'akov Kobi Gal. 2012. "Economic Games on the Internet: The Effect of $1 Stakes." *PLoS ONE* 7 (2). https://doi.org/10.1371/journal.pone.0031461

Andersen, Ross. 2017. "Welcome to Pleistocene Park." *The Atlantic*, April 2017.

Andreoni, James. 1995. "Warm-Glow versus Cold-Prickle: The Effects of Positive and Negative Framing on Cooperation in Experiments." *Quarterly Journal of Economics* 110 (1): 1–21. https://doi.org/10.2307/2118508

Andreoni, James, and John H. Miller. 1993. "Rational Cooperation in the Finitely Repeated Prisoner's Dilemma: Experimental Evidence." *Economic Journal* 103 (418): 570. https://doi.org/10.2307/2234532

Andrews, Talbot M., Andrew W. Delton, and Reuben Kline. 2018. "High-Risk High-Reward Investments to Mitigate Climate Change." *Nature Climate Change* 8 (September): 890–94. https://doi.org/10.1038/s41558-018-0266-y

Andrews, Talbot M., Andrew W. Delton, and Reuben Kline. 2021. "Is a Rational Politics of Disaster Possible? Making Useful Decisions for Others in an Experimental Disaster Game." *Political Behavior*, March, 1–22. https://doi.org/10.1007/s11109-021-09700-2

Andrews, Talbot M., Andrew W. Delton, and Reuben Kline. 2022a. "Anticipating Moral Hazard Undermines Climate Mitigation in an Experimental Geoengineering Game." *Ecological Economics* 196: 107421.

Andrews, Talbot M., Andrew W. Delton, and Reuben Kline. 2022b. "Who Do You Trust? Institutions That Constrain Leaders Help People Prevent Disaster." *Journal of Politics* 85 (1): 64–75.

Andrews, Talbot M., Reuben Kline, Yanna Krupnikov, and John Barry Ryan. 2022. "Too Many Ways to Help: How to Promote Climate Change Mitigation Behaviors." *Journal of Environmental Psychology* 81: 101806. https://doi.org/10.1016/j.jenvp.2022.101806

Andrews, Talbot M., and John Barry Ryan. 2021. "Preferences for Prevention: People Assume Expensive Problems Have Expensive Solutions." *Risk Analysis*, May, 13754. https://doi.org/10.1111/risa.13754

Andrews, Talbot M., Nicholas P. Simpson, Katharine J. Mach, and Christopher H. Trisos. 2023. "Risk from Response to a Changing Climate." *Climate Risk Management* 39: 100487.

Andrews, Talbot M., and Oleg Smirnov. 2020. "Who Feels the Impacts of Climate Change?" *Global Environmental Change* 65: 102164. https://doi.org/10.1016/j.gloenvcha.2020.102164

Atkeson, Lonna Rae, and Cherie D. Maestas. 2012. *Catastrophic Politics: How Extraordinary Events Redefine Perceptions of Government*. Cambridge University Press.

Axelrod, Lawrence J., and Darrin R. Lehman. 1993. "Responding to Environmental Concerns: What Factors Guide Individual Action?" *Journal of Environmental Psychology* 13 (2): 149–59. https://doi.org/10.1016/S0272-4944(05)80147-1

Bailey, Ronald. 2018. "Washington State Carbon Tax Initiative Loses." *Reason*, November 7, 2018.

Baker, A. C. 2001. "Reef Corals Bleach to Survive Change." *Nature* 411 (6839): 765–66. https://doi.org/10.1038/35081151

Baker, Mike. 2021. "Air-Conditioning Was Once Taboo in Seattle. Not Anymore." *New York Times*, June 25, 2021.

Baker, Mike, and Sergio Olmos. 2021. "The Pacific Northwest, Built for Mild Summers, Is Scorching Yet Again." *New York Times*, August 13, 2021.

Balliet, Daniel. 2010. "Communication and Cooperation in Social Dilemmas: A Meta-Analytic Review." *Journal of Conflict Resolution* 54 (1): 39–57. https://doi.org/10.1177/0022002709352443

Balliet, Daniel, and Paul A. M. Van Lange. 2013. "Trust, Punishment, and Cooperation Across 18 Societies." *Perspectives on Psychological Science* 8 (4): 363–79. https://doi.org/10.1177/1745691613488533

Barclay, Pat, and Robb Willer. 2007. "Partner Choice Creates Competitive Altruism in Humans." *Proceedings of the Royal Society B* 274 (1610): 749–53. https://doi.org/10.1098/rspb.2006.0209

Bar-Hillel, Maya, and Willem A Wagenaar. 1991. "The Perception of Randomness." *Advances in Applied Mathematics* 12 (4): 428–54. https://doi.org/10.1016/0196-8858(91)90029-I

Barnard, Anne. 2021. "The $119 Billion Sea Wall That Could Defend New York . . . or Not." *New York Times*, August 21, 2021.

Barrett, Scott, and Astrid Dannenberg. 2012. "Climate Negotiations under Scientific Uncertainty." *Proceedings of the National Academy of Sciences of the United States of America* 109 (43): 17372–76. https://doi.org/10.1073/pnas.1208417109

Barrett, Scott, and Astrid Dannenberg. 2014. "Sensitivity of Collective Action to Uncertainty about Climate Tipping Points." *Nature Climate Change* 4 (1): 36–39. https://doi.org/10.1038/nclimate2059

Bartels, Larry M. 2002. "Beyond the Running Tally: Partisan Bias in Political Perceptions." *Political Behavior* 24 (2): 117–50. https://doi.org/10.1023/A:1021226224601

Bateson, Melissa. 2002. "Recent Advances in Our Understanding of Risk-Sensitive Foraging Preferences." *Proceedings of the Nutrition Society* 61 (4): 509–16. https://doi.org/10.1079/pns2002181

Batson, C. Dani, K. O'Quin, J. Fultz, M. Vanderplas, and A. M. Isen. 1983. "Influence of Self-Reported Distress and Empathy on Egoistic versus Altruistic Motivation to Help." *Journal of Personality and Social Psychology* 45 (3): 706–18. https://doi.org/10.1037/0022-3514.45.3.706

Bayulgen, Oksan, Carol Atkinson-Palombo, Mary Buchanan, and Lyle Scruggs. 2021. "Tilting at Windmills? Electoral Repercussions of Wind Turbine Projects in Minnesota." *Energy Policy* 159 (December): 112636. https://doi.org/10.1016/J.ENPOL.2021.112636

Becker, C. Dustin, and Elinor Ostrom. 2003. "Human Ecology and Resource Sustainability: The Importance of Institutional Diversity." *Annual Review of Ecology and Systematics* 26 (1): 113–33. https://doi.org/10.1146/ANNUREV.ES.26.110195.000553

Berinsky, Adam J., Gregory A. Huber, and Gabriel S. Lenz. 2012. "Evaluating Online Labor Markets for Experimental Research: Amazon.Com's Mechanical Turk." *Political Analysis* 20 (03): 351–68. https://doi.org/10.1093/pan/mpr057

Berl, Janet E., Richard D. McKelvey, Peter C. Ordeshook, and Mark D. Winer. 1976. "An Experimental Test of the Core in a Simple N-Person Cooperative Nonsidepayment Game." *Journal of Conflict Resolution*. https://doi.org/10.1177/002200277602000304

Bernton, Hal. 2018a. "If Washington Voters OK a Carbon-Pollution Fee, Who Decides How to Spend All That Money?" *Seattle Times*, September 16, 2018.

Bernton, Hal. 2018b. "Washington State Voters Reject Carbon-Fee Initiative." *Seattle Times*, November 6, 2018.

Bhatia, Kieran T., Gabriel A. Vecchi, Thomas R. Knutson, Hiroyuki Murakami, James Kossin, Keith W. Dixon, and Carolyn E. Whitlock. 2019. "Recent Increases in Tropical Cyclone Intensification Rates." *Nature Communications* 10 (1). https://doi.org/10.1038/s41467-019-08471-z

Biermann, Frank, Jeroen Oomen, Aarti Gupta, Saleem H. Ali, Ken Conca, Maarten A. Hajer, Prakash Kashwan, et al. 2022. "Solar Geoengineering: The Case for an International Non-Use Agreement." *Wiley Interdisciplinary Reviews: Climate Change*, e754. https://doi.org/10.1002/WCC.754

Blake, Eric S., Todd B. Kimberlain, Robert J. Berg, John P. Cangialosi, and John L. Beven. 2012. "Tropical Cyclone Report: Hurricane Sandy." National Oceanic and Atmospheric Administration (NOAA). https://www.nhc.noaa.gov/data/tcr/AL182012_Sandy.pdf

Boehm, Christopher. 2009. *Hierarchy in the Forest: The Evolution of Egalitarian Behavior*. Harvard University Press.

Böhm, Robert, Özgür Gürerk, and Thomas Lauer. 2020. "Nudging Climate Change Mitigation: A Laboratory Experiment with Inter-Generational Public Goods." *Games* 11 (4): 1–20. https://doi.org/10.3390/g11040042

Bolsen, Toby, and James N. Druckman. 2018. "Do Partisanship and Politicization Undermine the Impact of a Scientific Consensus Message about Climate Change?" *Group Processes & Intergroup Relations* 21 (3): 389–402. https://doi.org/10.1177/1368430217737855

Boudet, Hilary, Leanne Giordono, Chad Zanocco, Hannah Satein, and Hannah Whitley. 2020. "Event Attribution and Partisanship Shape Local Discussion of Climate Change after Extreme Weather." *Nature Climate Change* 10 (1): 69–76. https://doi.org/10.1038/s41558-019-0641-3

Boudreau, Cheryl. 2009. "Closing the Gap: When Do Cues Eliminate Differences between Sophisticated and Unsophisticated Citizens?" *Journal of Politics* 71 (3): 964–76. https://doi.org/10.1017/S0022381609090823

Box, George E. 1979. "Robustness in the Strategy of Scientific Model Building." In *Robustness in Statistics*, edited by R. L. Laurner and G. N. Wilkinson, 201–36. Academic Press.

Brick, Cameron, David K. Sherman, and Heejung S. Kim. 2017. "'Green to Be Seen' and 'Brown to Keep Down': Visibility Moderates the Effect of Identity on Pro-Environmental Behavior." *Journal of Environmental Psychology* 51 (August): 226–38. https://doi.org/10.1016/j.jenvp.2017.04.004

Brown, Paul. 2019. "Carbon Capture Could Save the Planet." *TruthDig*, October 31, 2019.

Brown, Thomas C., and Stephan Kroll. 2017. "Avoiding an Uncertain Catastrophe: Climate Change Mitigation under Risk and Wealth Heterogeneity." *Climatic Change* 141 (2): 155–66.

Brulle, Robert J., Jason Carmichael, and J. Craig Jenkins. 2012. "Shifting Public Opinion on Climate Change: An Empirical Assessment of Factors Influencing Concern over Climate Change in the U.S., 2002–2010." *Climatic Change* 114 (2): 169–88. https://doi.org/10.1007/s10584-012-0403-y

Buchanan, James M., and G. Tullock. 1962. *The Calculus of Consent*. 3rd ed. University of Michigan Press.

Buhrmester, Michael, Tracy Kwang, and Samuel D. Gosling. 2011. "Amazon's Mechanical Turk: A New Source of Inexpensive, Yet High-Quality, Data?" *Perspectives on Psychological Science*. https://doi.org/10.1177/1745691610393980

Bump, Philip. 2015. "Jim Inhofe's Snowball Has Disproven Climate Change Once and for All." *Washington Post*, February 26, 2015.

Burton-Chellew, Maxwell N., Robert M. May, and Stuart A. West. 2013. "Combined Inequality in Wealth and Risk Leads to Disaster in the Climate Change Game." *Climatic Change* 120 (4): 815–30. https://doi.org/10.1007/s10584-013-0856-7

Butler, Daniel M., and David W. Nickerson. 2011. "Can Learning Constituency Opinion Affect How Legislators Vote? Results from a Field Experiment." *Quarterly Journal of Political Science* 6 (1): 55–83.

Camerer, Colin F. 2011. *Behavioral Game Theory: Experiments in Strategic Interaction*. Princeton University Press.

Campbell, A, P. E. Converse, W. E. Miller, and D. E. Stokes. 1980. *The American Voter*. University of Chicago Press.

Caplan, Bryan. 2011. *The Myth of the Rational Voter: Why Democracies Choose Bad Policies*. Princeton University Press.

Caraco, Thomas, Steven Martindale, and Thomas S. Whittam. 1980. "An Empirical Demonstration of Risk-Sensitive Foraging Preferences." *Animal Behaviour* 28 (3): 820–30. https://doi.org/10.1016/S0003-3472(80)80142-4

Carmichael, Jason T., and Robert J. Brulle. 2017. "Elite Cues, Media Coverage, and Public Concern: An Integrated Path Analysis of Public Opinion on Climate Change, 2001–2013." *Environmental Politics* 26 (2): 232–52. https://doi.org/10.1080/09644016.2016.1263433

Carmichael, Jason T., Robert J. Brulle, and Joanna K. Huxster. 2017. "The Great Divide: Understanding the Role of Media and Other Drivers of the Partisan Divide in Public Concern over Climate Change in the USA, 2001–2014." *Climatic Change* 141 (4): 599–612. https://doi.org/10.1007/s10584-017-1908-1

Case, Charleen R., and Jon K. Maner. 2015. "When and Why Power Corrupts: An Evolutionary Perspective." In *Handbook on Evolution and Society: Toward and Evolutionary Social Science*, edited by R. Machalek, J. Turner, and A. Maryanski, 460–73. Paradigm Publishers.

Caviola, Lucius, Stefan Schubert, and Jason Nemirow. 2020. "The Many Obstacles to Effective Giving." *Judgement and Decision Making* 15 (2): 159–72.

Chandler, William, P.R. Shukla, Roberto Schaeffer, Zhou Dadi, Fernando Tudela, Ogunlade Davidson, and Sema Alpan-Atamer. 2002. "Climate Change Mitigation in Developing Countries."US Department of Energy Office of Scientific and Technical Information. https://www.osti.gov/biblio/20824026

Charness, G., and M. Rabin. 2002. "Understanding Social Preferences with Simple Tests." *Quarterly Journal of Economics* 117 (3): 817–69. https://doi.org/10.1162/003355302760193904

Chaudhuri, Ananish. 2011. "Sustaining Cooperation in Laboratory Public Goods Experiments: A Selective Survey of the Literature." *Experimental Economics* 14 (1): 47–83. https://doi.org/10.1007/s10683-010-9257-1

Chiu, Allyson. 2019. "A Senator's Argument Against the Green New Deal: A Machine Gun–Toting Ronald Reagan Riding a Veloceraptor." *Washington Post*, March 27, 2019.

Chu, Haoran, and Janet Z. Yang. 2020. "Risk or Efficacy? How Psychological Distance Influences Climate Change Engagement." *Risk Analysis* 40 (4): 758–70. https://doi.org/10.1111/RISA.13446

"CO2 Emissions (Metric Tons per Capita)." 2020. https://data.worldbank.org/indicator/EN.ATM.CO2E.PC

Cody, Emily M., Jennie C. Stephens, James P. Bagrow, Peter Sheridan Dodds, and Christopher M. Danforth. 2017. "Transitions in Climate and Energy Discourse between Hurricanes Katrina and Sandy." *Journal of Environmental Studies and Sciences* 7 (1): 87–101. https://doi.org/10.1007/s13412-016-0391-8

Colten, Craig E. 2006. "Vulnerability and Place: Flat Land and Uneven Risk in New Orleans." *American Anthropologist* 108 (4): 731–34.

Congress, U.S. 2006. "A Failure of Initiative: Final Report of the Select Bipartisan Committee to Investigate the Preparation for and Response to Hurricane Katrina."U.S. Government Printing Office.

Converse, Philip E. 1964. "The Nature of Belief Systems in Mass Publics." *Critical Review* 18 (1–3): 1–74. https://doi.org/10.1080/08913810608443650

Corner, Adam, Nick Pidgeon, and Karen Parkhill. 2012. "Perceptions of Geoengineering: Public Attitudes, Stakeholder Perspectives, and the Challenge of 'Upstream' Engagement." *Wiley Interdisciplinary Reviews: Climate Change* 3 (5): 451–66. https://doi.org/10.1002/wcc.176

Cornes, Richard, and Todd Sandler. 1996. *The Theory of Externalities, Public Goods, and Club Goods*. Cambridge University Press.

Cosmides, Leda, and John Tooby. 1996. "Are Humans Good Intuitive Statisticians after All? Rethinking Some Conclusions from the Literature on Judgment under Uncertainty." *Cognition* 58 (1): 1–73. https://doi.org/10.1016/0010-0277(95)00664-8

Cox, Michael, Gwen Arnold, and Sergio Villamayor Tomás. 2010. "A Review of Design Principles for Community-Based Natural Resource Management on JSTOR." *Ecology and Society* 15 (4).

Cranley, Ellen. 2019. "These Are the 130 Current Members of Congress Who Have Doubted or Denied Climate Change." *Business Insider*, April 29, 2019.

Dagon, Katherine, and Daniel P. Schrag. 2016. "Exploring the Effects of Solar Radiation Management on Water Cycling in a Coupled Land–Atmosphere Model." *Journal of Climate* 29 (7): 2635–50. https://doi.org/10.1175/JCLI-D-15-0472.1

Dash, Nicole, and Hugh Gladwin. 2007. "Evacuation Decision Making and Behavioral Responses: Individual and Household." *Natural Hazards Review* 8 (3): 69–77. https://doi.org/10.1061/(asce)1527-6988(2007)8:3(69)

Dawes, Christopher T., Peter John Loewen, and James H. Fowler. 2015. "Social Preferences and Political Participation." *Journal of Politics* 73 (3): 845–56. https://doi.org/10.1017/S0022381611000508

Dellavigna, Stefano, John A. List, Ulrike Malmendier, and Gautam Rao. 2017. "Voting to Tell Others." *Review of Economic Studies* 84 (1): 143–81. https://doi.org/10.1093/restud/rdw056

Delli Carpini, Michael X., and Scott Keeter. 1993. "Measuring Political Knowledge: Putting First Things First." *American Journal of Political Science* 37 (4): 1179. https://doi.org/10.2307/2111549

Del Ponte, Alessandro, Andrew W. Delton, Reuben Kline, and Nicholas A. Seltzer. 2017. "Passing It Along: Experiments on Creating the Negative Externalities of Climate Change." *Journal of Politics* 79 (4): 1444–48. https://doi.org/10.1086/692472

Delton, Andrew W., and Max M. Krasnow. 2014. "An Independent Replication That the Evolution of Direct Reciprocity under Uncertainty Explains One-Shot Cooperation: Commentary on Zefferman." *Evolution and Human Behavior* 35 (6): 547–48.

Delton, Andrew W., Max M. Krasnow, Leda Cosmides, and John Tooby. 2011a. "No Reply to McNally and Tanner: Generosity Evolves When Cooperative Decisions Must Be Made under Uncertainty." *PNAS* 108 (44): E972.

Delton, Andrew W., Max M. Krasnow, Leda Cosmides, and John Tooby. 2011b. "Evolution of Direct Reciprocity under Uncertainty Can Explain Human Generosity in One-Shot Encounters." *Proceedings of the National Academy of Sciences of the United States of America* 108 (32): 13335–40. https://doi.org/10.1073/pnas.1102131108

Delton, Andrew W., Jason Nemirow, Theresa E. Robertson, Also Cimino, and Leda Cosmides. 2013. "Merely Opting out of a Public Good Is Moralized: An Error Management Approach to Cooperation." *Journal of Personality and Social Psychology* 105 (4): 621.

Delton, Andrew W., Michael Bang Petersen, and Theresa E. Robertson. 2018. "Partisan Goals, Emotions, and Political Mobilization: The Role of Motivated Reasoning in Pressuring Others to Vote." *Journal of Politics* 80 (3): 890–902. https://doi.org/10.1086/697124

Delton, Andrew W., and Aaron Sell. 2014. "The Co-Evolution of Concepts and

Motivation." *Current Directions in Psychological Science* 23 (2): 115–20. https://doi.org/10.1177/0963721414521631

DeScioli, Peter, Bowen Cho, Scott Bokemper, and Andrew W. Delton. 2018. "Selfish and Cooperative Voting: Can the Majority Restrain Themselves?" *Political Behavior*, 2018. https://doi.org/10.1007/s11109-018-9495-z

DeScioli, Peter, and Siddhi Krishna. 2013. "Giving to Whom? Altruism in Different Types of Relationships." *Journal of Economic Psychology* 34 (February): 218–28. https://doi.org/10.1016/J.JOEP.2012.10.003

DeSombre, Elizabeth R. 2018. *Why Good People Do Bad Environmental Things.* Oxford University Press.

Deutchman, Paul, Dorsa Amir, Matthew R. Jordan, and Katherine McAuliffe. 2021. "Common Knowledge Promotes Cooperation in the Threshold Public Goods Game by Reducing Uncertainty." *Evolution and Human Behavior*, December. https://doi.org/10.1016/J.EVOLHUMBEHAV.2021.12.003

Dietz, Thomas, Elinor Ostrom, and Paul C Stern. 2003. "The Struggle to Govern the Commons." *Science* 302 (5652): 1907–12. https://doi.org/10.1126/science.1091015

Diffenbaugh, Noah S., Martin Scherer, and Robert J. Trapp. 2013. "Robust Increases in Severe Thunderstorm Environments in Response to Greenhouse Forcing." *PNAS* 110 (41): 16361–66. https://doi.org/10.1073/pnas.1307758110

Dillion, Jar, and Phil Cross. 2013. "Bond Money Misused on Rural Water Dam." *Fox 25*. June 18, 2013.

Doherty, Kathryn L., and Thomas N. Webler. 2016. "Social Norms and Efficacy Beliefs Drive the Alarmed Segment's Public-Sphere Climate Actions." *Nature Climate Change* 6 (9): 879–84. https://doi.org/10.1038/nclimate3025

Donner, Simon D., and Sophie Webber. 2014. "Obstacles to Climate Change Adaptation Decisions: A Case Study of Sea-Level Rise and Coastal Protection Measures in Kiribati." *Sustainability Science* 9 (3): 331–45. https://doi.org/10.1007/s11625-014-0242-z

Downs, Anthony. 1957. "An Economic Theory of Political Action in a Democracy." *Journal of Political Economy* 65 (2): 135–50. https://doi.org/10.1086/257897

Drake, Henri F., and Geoffrey Henderson. 2022. "A Defense of Usable Climate Mitigation Science: How Science Can Contribute to Social Movements." *Climatic Change* 172 (1–2): 10..

Edlin, Aaron, Andrew Gelman, and Noah Kaplan. 2007. "Voting as a Rational Choice." *Rationality and Society* 19 (3): 293–314. https://doi.org/10.1177/1043463107077384

Egan, Patrick J., and Megan Mullin. 2017. "Climate Change: US Public Opinion." *Annual Review of Political Science* 20: 209–36. https://doi.org/10.1146/annurev-polisci-051215-022857

Elsner, James B., Svetoslava C. Elsner, and Thomas H. Jagger. 2015. "The Increasing Efficiency of Tornado Days in the United States." *Climate Dynamics* 45 (3–4): 651–59. https://doi.org/10.1007/s00382-014-2277-3

Emanuel, Kerry. 2005. "Increasing Destructiveness of Tropical Cyclones over the Past 30 Years." *Nature* 436 (7051): 686–88. https://doi.org/10.1038/nature03906

Evans, Simon. 2021. "Analysis: Which Countries Are Historically Responsible for Climate Change?" *CarbonBrief*. October 5, 2021.

Factbook, CIA World. 2019. "Kiribati." Washington, DC.

Falconer, Bruce. 2018. "Can Anyone Stop the Man Who Will Try Just about Anything to Put an End to Climate Change?" *Pacific Standard*, May.

Feddersen, Timothy, Sean Gailmard, and Alvaro Sanfroni. 2009. "Moral Bias in Large Elections: Theory and Experimental Evidence." *American Political Science Review* 103 (2): 175–92.

Fehr, Ernst, and Simon Gächter. 2000. "Cooperation and Punishment in Public Goods Experiments." *American Economic Review* 90 (4): 980–94. https://doi.org/10.1257/aer.90.4.980

Fehr, Ernst, and Simon Gächter. 2002. "Altruistic Punishment in Humans." *Nature* 415 (6868): 137–40. https://doi.org/10.1038/415137a

Fehr, Ernst, and Klaus M. Schmidt. 1999. "A Theory of Fairness, Competition, and Cooperation." *Quarterly Journal of Economics* 114 (3): 817–68.

Feinberg, Matthew, and Robb Willer. 2011. "Apocalypse Soon?" *Psychological Science* 22 (1): 34–38. https://doi.org/10.1177/0956797610391911

Feldman, Lauren, P. Sol Hart, and Tijana Milosevic. 2017. "Polarizing News? Representations of Threat and Efficacy in Leading US Newspapers' Coverage of Climate Change." *Public Understanding of Science* 26 (4): 481–97. https://doi.org/10.1177/0963662515595348

Finkbeiner, E. M., F. Micheli, A. Saenz-Arroyo, L. Vazquez-Vera, C. A. Perafan, and J. C. Cárdenas. 2018. "Local Response to Global Uncertainty: Insights from Experimental Economics in Small-Scale Fisheries." *Global Environmental Change* 48 (January): 151–57. https://doi.org/10.1016/J.GLOENVCHA.2017.11.010

Finucane, Melissa L., Ali Alhakami, Paul Slovic, and Stephen M. Johnson. 2000. "The Affect Heuristic in Judgments of Risks and Benefits." *Journal of Behavioral Decision Making* 13 (1): 1–17. https://doi.org/10.1002/(SICI)1099-0771(200001/03)13:1<1::AID-BDM333>3.0.CO;2-S

Fischbacher, Urs, Simon Gächter, and Ernst Fehr. 2001. "Are People Conditionally Cooperative? Evidence from a Public Goods Experiment." *Economics Letters* 71 (3): 397–404. https://doi.org/10.1016/S0165-1765(01)00394-9

Fiske, Alan P. 1992. "The Four Elementary Forms of Sociality: Framework for a Unified Theory of Social Relations." *Psychological Review* 99 (4): 689–723. https://doi.org/10.1037/0033-295X.99.4.689

Fornasari, Federico, Matteo Ploner, and Ivan Soraperra. 2020. "Interpersonal Risk Assessment and Social Preferences: An Experimental Study." *Journal of Eco-

nomic Psychology 77 (March): 102183. https://doi.org/10.1016/j.joep.2019.06.006

Friedman, Jeffrey A. 2019. “Priorities for Preventive Action: Explaining Americans’ Divergent Reactions to 100 Public Risks.” *American Journal of Political Science* 63 (1): 181–96. https://doi.org/10.1111/ajps.12400

Friedman, Milton. 1953. “The Methodology of Positive Economics.” In *The Philosophy of Economics: An Anthology*, edited by Daniel M. Hausman. Cambridge University Press.

Funk, Cary, and Meg Hefferon. 2019. “U.S. Public Views on Climate and Energy.” Pew Research Center, November 25, 2019 https://www.pewresearch.org/science/2019/11/25/u-s-public-views-on-climate-and-energy/

Gailmard, Sean, and John W. Patty. 2018. “Preventing Prevention.” *American Journal of Political Science* 63 (2): 342–52.

Garfield, Zachary H., Christopher von Rueden, and Edward H. Hagen. 2019. “The Evolutionary Anthropology of Political Leadership.” *Leadership Quarterly* 30 (1): 59–80. https://doi.org/10.1016/J.LEAQUA.2018.09.001

Gifford, Robert. 2011. “The Dragons of Inaction: Psychological Barriers That Limit Climate Change Mitigation and Adaptation.” *American Psychologist* 66 (4): 290–302. https://doi.org/10.1037/a0023566

Gigerenzer, Gerd. 1998. “Ecological Intelligence: An Adaptation for Frequencies.” In *The Evolution of Mind*, edited by Denise Dellarosa Cummins and Colin Allen, 9–29. Oxford University Press.

Gilens, Martin. 2012. *Affluence and Influence: Economic Inequality and Political Power in America*. Princeton University Press.

Gleditsch, Nils Petter. 2012. “Whither the Weather? Climate Change and Conflict.” *Journal of Peace Research* 49 (1): 3–9. https://doi.org/10.1177/0022343311431288

Glimcher, Paul W. 2011. *Foundations of Neuroeconomic Analysis*. Oxford University Press.

González-Alemán, Juan J., Salvatore Pascale, Jesús Gutierrez-Fernandez, Hiroyuki Murakami, Miguel A. Gaertner, and Gabriel A. Vecchi. 2019. “Potential Increase in Hazard From Mediterranean Hurricane Activity with Global Warming.” *Geophysical Research Letters* 46 (3): 1754–64. https://doi.org/10.1029/2018GL081253

“Governor Kate Brown Signs Bill to Modernize and Improve Wildfire Preparedness.” July 30, 2021. State of Oregon Newsroom.

Gray, W. M. 2005. “The Role of Science in Environmental Policy Making.” https://www.epw.senate.gov/public/index.cfm/hearings?Id=E00DDE50-802A-23AD-436F-443DA217C01E&Statement_id=CD701C3D-4F5A-4147-B391-EB43C714C86D

Griskevicius, Vladas, Joshua M. Tybur, and Bram Van den Bergh. 2010. “Going Green to Be Seen: Status, Reputation, and Conspicuous Conservation.” *Journal of Personality and Social Psychology* 98 (3): 392–404. https://doi.org/10.1037/a0017346

Guess, Andrew M. 2021. "(Almost) Everything in Moderation: New Evidence on Americans' Online Media Diets." *American Journal of Political Science*. https://doi.org/10.1111/ajps.12589

Haidt, Jonathan. 2012. *The Righteous Mind: Why Good People Are Divided by Politics and Religion*. Vintage.

Hale, Benjamin. 2012. "The World That Would Have Been: Moral Hazard Arguments Against Geoengineering." In *Reflecting Sunlight: The Ethics of Solar Radiation Management*, edited by Christopher Preston. Rowman and Littlefield.

Hamman, John R., Roberto A. Weber, and Jonathan Woon. 2011. "An Experimental Investigation of Electoral Delegation and the Provision of Public Goods." *American Journal of Political Science* 55 (4): 738–52. https://doi.org/10.1111/j.1540-5907.2011.00531.x

Hansen, James, Makiko Sato, Pushker Kharecha, David Beerling, Robert Berner, Valerie Masson-Delmotte, Mark Pagani, Maureen Raymo, Dana L. Royer, and James C. Zachos. 2008. "Target Atmospheric CO: Where Should Humanity Aim?" *Open Atmospheric Science Journal* 2 (1): 217–31. https://doi.org/10.2174/1874282300802010217

Hansen, J., M. Sato, R. Ruedy, P. Kharecha, A. Lacis, R. Miller, L. Nazarenko, et al. 2007. "Dangerous Human-Made Interference with Climate: A GISS Model Study." *Atmospheric Chemistry and Physics* 7 (9): 2287–2312. https://doi.org/10.5194/acp-7-2287-2007

Hastie, Reid, and Robyn M. Dawes. 2009. *Rational Choice in an Uncertain World: The Psychology of Judgement and Decision Making*. Sage Publications.

Hauser, David J., and Norbert Schwarz. 2016. "Attentive Turkers: MTurk Participants Perform Better on Online Attention Checks than Do Subject Pool Participants." *Behavior Research Methods* 48 (1): 400–407. https://doi.org/10.3758/s13428-015-0578-z

Healy, Andrew, and Neil Malhotra. 2009. "Myopic Voters and Natural Disaster Policy." *American Political Science Review* 103 (03): 387–406. https://doi.org/10.1017/S0003055409990104

Henrich, Joseph, Robert Boyd, Samuel Bowles, Colin Camerer, Ernst Fehr, Herbert Gintis, Richard McElreath, et al. 2005. "'Economic Man' in Cross-Cultural Perspective: Behavioral Experiments in 15 Small-Scale Societies." *Behavioral and Brain Sciences* 28 (06): 795–815. https://doi.org/10.1017/S0140525X05000142

Hersher, Rebecca. 2020. "States Prepare to Spend Millions to Address Flooding." *NPR*, July 6, 2020.

Hertwig, Ralph, and Gerd Gigerenzer. 1999. "The 'Conjunction Fallacy' Revisited: How Intelligent Inferences Look like Reasoning Errors." *Behavioral Decision Making* 12 (4): 275–305.

Homer-Dixon, Thomas F. 1999. *Environment, Scarcity, and Violence*. Princeton University Press.

Howe, Peter D., Jennifer Marlon, Matto Mildenberger, and Brittany S. Shield.

2019. "How Will Climate Change Shape Climate Opinion?" *Environmental Research Letters*, September. https://doi.org/10.1088/1748-9326/ab466a

Huckfeldt, Robert. 2007. "Unanimity, Discord, and the Communication of Public Opinion." *American Journal of Political Science* 51 (4): 978–95. https://doi.org/10.1111/j.1540-5907.2007.00292.x

Huckfeldt, Robert, and Jeanette Morehouse Mendez. 2015. "Moths, Flames, and Political Engagement: Managing Disagreement within Communication Networks." *Journal of Politics*, July. https://doi.org/10.1017/S0022381607080073

Incropera, Frank P. 2016. *Climate Change: A Wicked Problem: Complexity and Uncertainty at the Intersection of Science, Economics, Politics, and Human Behavior*. Cambridge University Press.

Intelligence Squared. 2019. "Engineering Solar Radiation Is a Crazy Idea." 2019. https://www.intelligencesquaredus.org/debate/engineering-solar-radiation-crazy-idea/#/

IPCC. 2010. *Guidance Note for Lead Authors of the IPCC Fifth Assessment Report on Consistent Treatment of Uncertainties*. Jasper Ridge, CA.

IPCC. 2014a. *Climate Change 2014: Mitigation Report. Contribution of Working Group III to the Fifth Assessment Report on the Intergovernmental Panel on Climate Change*, edited by Rajendra K. Pachouri and Louis A. Meyer. Geneva, Switzerland.

IPCC. 2014b. *Climate Change 2014: Synthesis Report. Contribution of Working Groups I, II, and III to the Fifth Assessment Report of the Intergovernmental Panel on Climate Change*, edited by Rajendra K Pachauri and Louis A. Meyer. Geneva, Switzerland.

IPCC. 2018. *Global Warming of 1.5oC*. Geneva, Switzerland.

IPCC. 2021. *Climate Change 2021: The Physical Science Basis. Contribution of Working Group I to the Sixth Assessment Report of the Intergovernmental Panel on Climate Change*, edited by V. Masson-Delmotte, P. Zhai, A. Pirani, S. L. Connors, C. Péan, S. Berger, N. Caud, et al. Cambridge University Press.

Ives, Mike. 2016. "A Remote Pacific Nation, Threatened by Rising Seas." *New York Times*, July 2, 2016.

Jebari, Joseph, Olúfẹ́mi O. Táíwò, Talbot M. Andrews, Valentina Aquila, Brian Beckage, Mariia Belaia, Maggie Clifford, et al. 2021. "From Moral Hazard to Risk-Response Feedback." *Climate Risk Management* 33 (January): 100324. https://doi.org/10.1016/j.crm.2021.100324

Jézéquel, Aglaé, Vivian Dépoues, Hélène Guillemot, Amélie Rajaud, Mélodie Trolliet, Mathieu Vrac, Jean Paul Vanderlinden, and Pascal Yiou. 2020. "Singular Extreme Events and Their Attribution to Climate Change: A Climate Service–Centered Analysis." *Weather, Climate, and Society* 12 (1): 89–101. https://doi.org/10.1175/WCAS-D-19-0048.1

Jézéquel, Aglaé, Vivian Dépoues, Hélène Guillemot, Mélodie Trolliet, Jean Paul Vanderlinden, and Pascal Yiou. 2018. "Behind the Veil of Extreme Event Attribution." *Climatic Change* 149 (3–4): 367–83. https://doi.org/10.1007/s10584-018-2252-9

Jordan, Matthew R., Jillian J. Jordan, and David G. Rand. 2017. "No Unique Effect of Intergroup Competition on Cooperation: Non-Competitive Thresholds Are as Effective as Competitions between Groups for Increasing Human Cooperative Behavior." *Evolution and Human Behavior* 38 (1): 102–8. https://doi.org/10.1016/j.evolhumbehav.2016.07.005

Kahan, Dan M., Ellen Peters, Maggie Wittlin, Paul Slovic, Lisa Larrimore Ouellette, Donald Braman, and Gregory Mandel. 2012. "The Polarizing Impact of Science Literacy and Numeracy on Perceived Climate Change Risks." *Nature Climate Change* 2 (10): 732–35. https://doi.org/10.1038/nclimate1547

Kameda, Tatsuya, Masanori Takezawa, R. Scott Tindale, and Christine M. Smith. 2002. "Social Sharing and Risk Reduction Exploring a Computational Algorithm for the Psychology of Windfall Gains." *Evolution and Human Behavior* 23 (1): 11–33. https://doi.org/10.1016/S1090-5138(01)00086-1

Keith, David W. 2000. "Geoengineering the Climate: History and Prospect." *Annual Review of Energy and the Environment* 25 (1): 245–84. https://doi.org/10.1146/annurev.energy.25.1.245

Keohane, Robert O., and David G. Victor. 2016. "Cooperation and Discord in Global Climate Policy." *Nature Climate Change*. https://doi.org/10.1038/nclimate2937

KGW Staff. 2021. "Gov. Brown Signs $220 Million Wildfire Prevention Bill." KGW, July 30, 2021. https://www.kgw.com/article/news/local/wildfire/brown-signs-wildfire-prevention-bill/283-49261003-8e59-41fb-86e7-e98644aa4953

Kline, Reuben, Nicholas Seltzer, Evgeniya Lukinova, and Autumn Bynum. 2018. "Differentiated Responsibilities and Prosocial Behaviour in Climate Change Mitigation." *Nature Human Behaviour* 2 (9): 653–61. https://doi.org/10.1038/s41562-018-0418-0

Kocher, Martin G., Todd Cherry, Stephan Kroll, Robert J. Netzer, and Matthias Sutter. 2008. "Conditional Cooperation on Three Continents." *Economics Letters* 101 (3): 175–78. https://doi.org/10.1016/J.ECONLET.2008.07.015

Kuklinski, James H., and Paul J. Quirk. 2000. "Reconsidering the Rational Public: Cognition, Heuristics, and Mass Opinion." In *Elements of Reason: Cognition, Choice, and the Bounds of Rationality*, edited by Arthur Lupia, Mathew D. McCubbins, and Samuel L. Popkin, 951–71. Cambridge University Press.

Kurzban, Robert, and Daniel Houser. 2005. "Experiments Investigating Cooperative Types in Humans: A Complement to Evolutionary Theory and Simulations." *PNAS* 102 (5): 1803–7. https://doi.org/10.1073/pnas.0408759102

Lax, Jeffer R., and Justin H. P Phillips. 2009. "Gay Rights in the States: Public Opinion and Policy Responsiveness." *American Political Science Review* 103 (3): 367–86. https://doi.org/10.1017/S0003055409990050

Leber, Rebecca. 2021. "ExxonMobil Wants You to Feel Responsible for Climate Change So It Doesn't Have To." *Vox*, May 13, 2021.

Ledyard, John O. 1995. "Public Goods: A Survey of Experimental Research." In *The Handbook of Experimental Economics*, edited by John H. Kagel and Alvin E. Roth, 111–94. Princeton University Press.

Ledyard, John O., and Kristin Szakaly-Moore. 1994. "Designing Organizations for

Trading Pollution Rights." *Journal of Economic Behavior & Organization* 25 (2): 167–96. https://doi.org/10.1016/0167-2681(94)90009-4

Leiserowitz, Anthony. 2005. "American Risk Perceptions: Is Climate Change Dangerous?" *Risk Analysis* 25 (6): 1433–42. https://doi.org/10.1111/j.1540-6261.2005.00690.x

Leiserowitz, A., E. Maibach, S. Rosenthal, J. Kotcher, M. Ballew, M. Goldberg, and A. Gustafson. 2019. "Climate Change in the American Mind: November 2019." New Haven, CT. Yale Program on Climate Communication Report, https://climatecommunication.yale.edu/publications/climate-change-in-the-american-mind-november-2019/

Lenton, Timothy M. 2011. "Early Warning of Climate Tipping Points." *Nature Climate Change* 1 (4): 201–9. https://doi.org/10.1038/nclimate1143

Lenton, Timothy M., Hermann Held, Elmar Kriegler, Jim W. Hall, Wolfgang Lucht, Stefan Rahmstorf, and Hans Joachim Schellnhuber. 2008. "Tipping Elements in the Earth's Climate System." *PNAS*, https://doi.org/10.1073/pnas.0705414105

LeVeck, Brad L., D. Alex Hughes, James H. Fowler, Emilie Hafner-Burton, and David G. Victor. 2014. "The Role of Self-Interest in Elite Bargaining." *PNAS* 111 (52): 18536–41. https://doi.org/10.1073/pnas.1409885111

Levine, Adam Seth, and Reuben Kline. 2017. "A New Approach for Evaluating Climate Change Communication." *Climatic Change* 142 (1–2): 301–9. https://doi.org/10.1007/s10584-017-1952-x

Lin, Albert C. 2013. "Does Geoengineering Present a Moral Hazard?" *Ecology Law Quarterly* 40 (3): 673–712. https://doi.org/10.15779/Z38JP1J

Lindell, Michael K., and Ronald W. Perry. 2012. "The Protective Action Decision Model: Theoretical Modifications and Additional Evidence." *Risk Analysis* 32 (4): 616–32. https://doi.org/10.1111/j.1539-6924.2011.01647.x

Lindsey, Rebecca. 2020. "Climate Change: Global Sea Level." NOAA: Climate.gov. https://www.climate.gov/news-features/understanding-climate/climate-change-global-sea-level

Lindsey, Rebecca. 2019. "Climate Change: Atmospheric Carbon Dioxide." NOAA: Climate.Gov. https://www.climate.gov/news-features/understanding-climate/climate-change-atmospheric-carbon-dioxide

Live, Oregon. 2021. "Portland's June 2021 Heat Wave: What to Know." *The Oregonian*, June 29, 2021.

Lupia, Arthur, and Matthew D. McCubbins. 1998. *The Democratic Dilemma: Can Citizens Learn What They Need to Know?* Cambridge University Press.

Mahajan, Aseem, Dustin Tingley, and Gernot Wagner. 2019. "Fast, Cheap, and Imperfect? US Public Opinion about Solar Geoengineering." *Environmental Politics* 28 (3): 523–43. https://doi.org/10.1080/09644016.2018.1479101

Matthews, Dylan. 2017. "Donald Trump Has Tweeted Climate Change Skepticism 115 Times. Here's All of It." *Vox*, June 1, 2017.

McNally, Luke, and Colby J. Tanner. 2011. "Flexible Strategies, Forgiveness, and the Evolution of Generosity in One-Shot Encounters." *PNAS* 108 (44): E971. https://doi.org/10.1073/PNAS.1115225108

Merk, Christine, Gert Pönitzsch, and Katrin Rehdanz. 2016. "Knowledge about Aerosol Injection Does Not Reduce Individual Mitigation Efforts." *Environmental Research Letters* 11 (5): 054009. https://doi.org/10.1088/1748-9326/11/5/054009

Merkley, Eric, and Dominik A. Stecula. 2021. "Party Cues in the News: Democratic Elites, Republican Backlash, and the Dynamics of Climate Skepticism." *British Journal of Political Science* 51 (4): 1439–56. https://doi.org/10.1017/S0007123420000113

Mildenberger, Matto, Erick Lachapelle, Kathryn Harrison, and Isabelle Stadelmann-Steffen. 2022. "Limited Impacts of Carbon Tax Rebate Programmes on Public Support for Carbon Pricing." *Nature Climate Change 2022*, January, 1–7. https://doi.org/10.1038/s41558-021-01268-3

Milinski, Manfred, Christian Hilbe, Dirk Semmann, Ralf Sommerfeld, and Jochem Marotzke. 2016. "Humans Choose Representatives Who Enforce Cooperation in Social Dilemmas through Extortion." *Nature Communications* 7 (March): 10915. https://doi.org/10.1038/ncomms10915

Milinski, Manfred, Torsten Röhl, and Jochem Marotzke. 2011. "Cooperative Interaction of Rich and Poor Can Be Catalyzed by Intermediate Climate Targets." *Climatic Change* 109 (3–4): 807–14. https://doi.org/10.1007/s10584-011-0319-y

Milinski, Manfred, Ralf D. Sommerfeld, Hans-Jürgen Krambeck, Floyd A. Reed, and Jochem Marotzke. 2008. "The Collective-Risk Social Dilemma and the Prevention of Simulated Dangerous Climate Change." *PNAS* 105 (7): 2291–94. https://doi.org/10.1073/pnas.0709546105

Miller, Gary, and Thomas Hammond. 1994. "Why Politics Is More Fundamental than Economics." *Journal of Theoretical Politics* 6 (1): 5–26. https://doi.org/10.1177/0951692894006001001

Mishra, Sandeep. 2014. "Decision-Making Under Risk." *Personality and Social Psychology Review* 18 (3): 280–307. https://doi.org/10.1177/1088868314530517

Molnár, Péter K., Cecilia M. Bitz, Marika M. Holland, Jennifer E. Kay, Stephanie R. Penk, and Steven C. Amstrup. 2020. "Fasting Season Length Sets Temporal Limits for Global Polar Bear Persistence." *Nature Climate Change* 10 (8): 732–38. https://doi.org/10.1038/s41558-020-0818-9

Morgenstern, Oskar, and John Von Neumann. 1944. *Theory of Games and Economic Behavior*. Princeton University Press.

Morton, Rebecca B., and Kenneth C. Williams. 2010. *Experimental Political Science and the Study of Causality: From Nature to the Lab*. Cambridge University Press.

Motta, Matthew, and Andrew Rohrman. 2019. "Quaking in Their Boots? Inaccurate Perceptions of Seismic Hazard and Public Policy Inaction." *Behavioural Public Policy*, July, 1–17. https://doi.org/10.1017/bpp.2019.18

Muller, R. Andrew, and Stuart Mestelman. 1998. "What Have We Learned from

Emissions Trading Experiments?" *Managerial and Decision Economics*. Wiley. https://doi.org/10.1002/(sici)1099-1468(199806/08)19:4/5<225::aid-mde888>3.0.co;2-v

Nel, Philip, and Marjolein Righarts. 2008. "Natural Disasters and the Risk of Violent Civil Conflict." *International Studies Quarterly* 52 (1): 159–85. https://doi.org/10.1111/j.1468-2478.2007.00495.x

Nickerson, Raymond S. 2002. "The Production and Perception of Randomness." *Psychological Review* 109 (2): 330–57.

Noll, Roger G. 1982. "Implementing Marketable Emissions Permits." *American Economic Review* 72 (2): 120–24.

Nolt, John. 2019. "Domination across Space and Time: Smallpox, Relativity, and Climate Ethics." *Ethics, Policy and Environment* 22 (2): 172–83. https://doi.org/10.1080/21550085.2019.1625542

Nordhaus, William D. 2013. *The Climate Casino*. Yale University Press.

O'Connor, Robert E., Richard J. Bord, and Ann Fisher. 1999. "Risk Perceptions, General Environmental Beliefs, and Willingness to Address Climate Change." *Risk Analysis* 19 (3): 461–71. https://doi.org/10.1111/j.1539-6924.1999.tb00421.x

Okutsu, Akane, Erwida Maulia, and Apornrath Phoonphongphiphat. 2021. "The Climate Moonshot: Engineering the Earth." *Nikkei Asia*, November 2, 2021.

O'Neill, Brian C., and Michael Oppenheimer. 2002. "Dangerous Climate Impacts and the Kyoto Protocol." *Science*. https://doi.org/10.1126/science.1071238

Oppenheimer, M., and R. B. Alley. 2004. "The West Antarctic Ice Sheet and Long Term Climate Policy: An Editorial Comment." *Climatic Change*, May 2004. https://doi.org/10.1023/B:CLIM.0000024792.06802.31

Oppenheimer, Michael, and R. B. Alley. 2005. "Ice Sheets, Global Warming, and Article 2 of the UNFCCC: An Editorial Essay." *Climatic Change*. https://doi.org/10.1007/s10584-005-5372-y

Ostrom, Elinor. 1998. "A Behavioral Approach to the Rational Choice Theory of Collective Action: Presidential Address, American Political Science Association, 1997." *American Political Science Review* 92 (01): 1–22. https://doi.org/10.2307/2585925

Ostrom, Elinor. 2015. *Governing the Commons*. Cambridge University Press.

Ostrom, Elinor, Roy Gardner, and James M. Walker. 1994. *Rules, Games, and Common-Pool Resources*. University of Michigan Press.

Ostrom, Elinor, James Walker, and Roy Gardner. 1992. "Covenants with and without a Sword: Self-Governance Is Possible." *American Political Science Review* 86 (02): 404–17. https://doi.org/10.2307/1964229

Palomo-Vélez, Gonzalo, Joshua M. Tybur, and Mark van Vugt. 2021. "Is Green the New Sexy? Romantic of Conspicuous Conservation." *Journal of Environmental Psychology* 73: 101530. https://doi.org/10.1016/j.jenvp.2020.101530

Pauw, Pieter, Steffen Bauer, Carmen Richerzhagen, Clara Brandi, and Hanna Schmole. 2014. "Different Perspectives on Differentiated Responsibilities: A

State-of-the-Art Review of the Notion of Common but Differentiated Responsibilities in International Negotiations." Discussion paper, IDOS, German Institute of Development and Sustainability.

Petersen, Michael Bang, and Lene Aarøe. 2013. "Politics in the Mind's Eye: Imagination as a Link between Social and Political Cognition." *American Political Science Review* 107 (2): 275–93. https://doi.org/10.1017/S0003055413000026

Pidgeon, N., A. Corner, K. Parkhill, A. Spence, C. Butler, and W. Poortinga. 2012. "Exploring Early Public Responses to Geoengineering." *Philosophical Transactions of the Royal Society A: Mathematical, Physical and Engineering Sciences* 370 (1974): 4176–96. https://doi.org/10.1098/rsta.2012.0099

Pligt, Joop. 1998. "Perceived Risk and Vulnerability as Predictors of Precautionary Behaviour." *British Journal of Health Psychology* 3 (1): 1–14. https://doi.org/10.1111/j.2044-8287.1998.tb00551.x

Polman, Evan, and Kaiyang Wu. 2019. "Decision Making for Others Involving Risk: A Review and Meta-Analysis." *Journal of Economic Psychology*. Elsevier B.V. https://doi.org/10.1016/j.joep.2019.03.007

Poundstone, William. 1992. *Prisoner's Dilemma*. Anchor Books.

Price, Michael E., and Mark Van Vugt. 2014. "The Evolution of Leader-Follower Reciprocity: The Theory of Service-for-Prestige." *Frontiers in Human Neuroscience* 8: 363. https://doi.org/10.3389/FNHUM.2014.00363

Prior, Markus. 2005. "News vs. Entertainment: How Increasing Media Choice Widens Gaps in Political Knowledge and Turnout." *American Journal of Political Science* 49 (3): 577–92. https://doi.org/10.1111/j.1540-5907.2005.00143.x

Rabin, Matthew. 1993. "Incorporating Fairness into Game Theory and Economics." *American Economic Review* 5: 1281–302.

Raihani, Nichola. 2021. *The Social Instinct: How Cooperation Shaped the World*. St. Martin's Press.

Raimi, Kaitlin T., Alexander Maki, David Dana, and Michael P. Vandenbergh. 2019. "Framing of Geoengineering Affects Support for Climate Change Mitigation." *Environmental Communication* 13 (3): 300–319. https://doi.org/10.1080/17524032.2019.1575258

Ray, Aaron, Llewelyn Hughes, David M. Konisky, and Charles Kaylor. 2017. "Extreme Weather Exposure and Support for Climate Change Adaptation." *Global Environmental Change* 46: 104–13. https://doi.org/10.1016/j.gloenvcha.2017.07.002

Reeves, Andrew. 2011. "Political Disaster: Unilateral Powers, Electoral Incentives, and Presidential Disaster Declarations." *Journal of Politics* 73 (4): 1142–51. https://doi.org/10.1017/S0022381611000843

Regan, Helen. 2022. "Tonga Remains Cut off after a Massive Eruption and Tsunami. Here's What We Know." *CNN*, January 18, 2022.

Revelle, R. R. 1983. "Probable Future Changes in Sea Level Resulting from Increased Atmospheric Carbon Dioxide." In *Changing Climate*, 441–48. National Academy Press.

Riker, William H., and Peter C. Ordeshook. 1968. "A Theory of the Calculus of Voting." *American Political Science Review* 62 (01): 25–42. https://doi.org/10.1017/S000305540011562X

Roberts, David. 2018a. "Reckoning with Climate Change Will Demand Ugly Tradeoffs from Environmentalists—and Everyone Else." *Vox*, January 27, 2018.

Roberts, David. 2018b. "Washington Votes No on a Carbon Tax—Again." *Vox*, November 6, 2018.

Robinson, Kim Stanley. 2020. *The Ministry for the Future: A Novel.* Hatchett.

Rode, Catrin, Leda Cosmides, Wolfgang Hell, and John Tooby. 1999. "When and Why Do People Avoid Unknown Probabilities in Decisions under Uncertainty? Testing Some Predictions from Optimal Foraging Theory." *Cognition* 72 (3): 269–304. https://doi.org/10.1016/S0010-0277(99)00041-4

Roe, Gerard H, and Marcia B Baker. 2007. "Why Is Climate Sensitivity so Unpredictable?" *Science* 318 (5850): 629–32. https://doi.org/10.1126/science.1144735

Rolfe, Meredith. 2012. *Voter Turnout: A Social Theory of Political Participation.* Cambridge University Press.

Roxburgh, Nicholas, Dabo Guan, Kong Joo Shin, William Rand, Shunsuke Managi, Robin Lovelace, and Jing Meng. 2019. "Characterising Climate Change Discourse on Social Media during Extreme Weather Events." *Global Environmental Change* 54 (January): 50–60. https://doi.org/10.1016/j.gloenvcha.2018.11.004

Rozenberg, Julie, and Stéphane Hallegatte. 2018. "No Poor People on the Front Line: The Impacts of Climate Change on Poverty in 2030." In *Climate Justice: Integrating Economics and Philosophy*, edited by Ravi Kanbur and Henry Shue. Oxford University Press.

Rudman, Laurie A., Meghan C. McLean, and Martin Bunzl. 2013. "When Truth Is Personally Inconvenient, Attitudes Change: The Impact of Extreme Weather on Implicit Support for Green Politicians and Explicit Climate-Change Beliefs." *Psychological Science* 24 (11): 2290–96. https://doi.org/10.1177/0956797613492775

Sainz-Santamaria, Jaime, and Sarah E. Anderson. 2013. "The Electoral Politics of Disaster Preparedness." *Risk, Hazards & Crisis in Public Policy* 4 (4): 234–49. https://doi.org/10.1002/rhc3.12044

Sally, David. 1995. "Conversation and Cooperation in Social Dilemmas: A Meta-Analysis of Experiments from 1958 to 1992." *Rationality and Society* 9 (1). https://doi.org/10.1177/104346319500700100 4

Scannell, Leila, and Robert Gifford. 2013. "Personally Relevant Climate Change." *Environment and Behavior* 45 (1): 60–85. https://doi.org/10.1177/0013916511421196

Schelling, Thomas C. 1960. *The Strategy of Conflict.* Harvard University Press.

Scheuch, Eric. 2020. "Life after Coal: The Decline and Rise of West Virginia Coal Country." *State of the Planet*, August 7, 2020.

Scruggs, Lyle, and Salil Benegal. 2012. "Declining Public Concern about Climate

Change: Can We Blame the Great Recession?" *Global Environmental Change* 22 (2): 505–15. https://doi.org/10.1016/j.gloenvcha.2012.01.002

Shadish, William R., Thomas D. Cook, and Donald T. Campbell. 2002. *Experimental and Quasi-Experimental Designs for Generalized Causal Inference*. Wadsworth Cengage Learning.

Shao, Wanyun, and Kirby Goidel. 2016. "Seeing Is Believing? An Examination of Perceptions of Local Weather Conditions and Climate Change Among Residents in the U.S. Gulf Coast." *Risk Analysis* 36 (11): 2136–57. https://doi.org/10.1111/risa.12571

Sheffer, Lior, Peter John Loewen, Stuart Soroka, Stefaan Walgrave, and Tamir Sheafer. 2018. "Nonrepresentative Representatives: An Experimental Study of the Decision Making of Elected Politicians." *American Political Science Review* 112 (2): 302–21. https://doi.org/10.1017/S0003055417000569

Shepard, Stephanie, Hilary Boudet, Chad M. Zanocco, Lori A. Cramer, and Bryan Tilt. 2018. "Community Climate Change Beliefs, Awareness, and Actions in the Wake of the September 2013 Flooding in Boulder County, Colorado." *Journal of Environmental Studies and Sciences* 8 (3): 312–25. https://doi.org/10.1007/s13412-018-0479-4

Sherstyuk, Katerina, Nori Tarui, Majah-Leah V. Ravago, and Tatsuyoshi Saijo. 2016. "Intergenerational Games with Dynamic Externalities and Climate Change Experiments." *Journal of the Association of Environmental and Resource Economists* 3 (2): 247–81. https://doi.org/10.1086/684162

Simpson, I. R., S. Tilmes, J. H. Richter, B. Kravitz, D. G. MacMartin, M. J. Mills, J. T. Fasullo, and A. G. Pendergrass. 2019. "The Regional Hydroclimate Response to Stratospheric Sulfate Geoengineering and the Role of Stratospheric Heating." *Journal of Geophysical Research: Atmospheres* 124 (23): 12587–616. https://doi.org/10.1029/2019JD031093

Simpson, Nicholas P., Talbot M. Andrews, Matthias Krönke, Christopher Lennard, Romaric C. Odoulami, Birgitt Ouweneel, Anna Steynor, and Christopher H. Trisos. 2021. "Climate Change Literacy in Africa." *Nature Climate Change* 11 (11): 937–44. https://doi.org/10.1038/s41558-021-01171-x

Slettebak, Rune T. 2012. "Don't Blame the Weather! Climate-Related Natural Disasters and Civil Conflict." *Journal of Peace Research* 49 (1): 163–76. https://doi.org/10.1177/0022343311425693

Slovic, Paul, Melissa L. Finucane, Ellen Peters, and Donald G. MacGregor. 2004. "Risk as Analysis and Risk as Feelings: Some Thoughts about Affect, Reason, Risk, and Rationality." *Risk Analysis*. John Wiley & Sons. https://doi.org/10.1111/j.0272-4332.2004.00433.x

Smirnov, Oleg. 2019. "Collective Risk Social Dilemma and the Consequences of the US Withdrawal from International Climate Negotiations." *Journal of Theoretical Politics*, September, 095162981987551. https://doi.org/10.1177/0951629819875511

Smith, Vernon L. 1982. "Microeconomic Systems as Experimental Science." *American Economic Review* 72 (5): 923–55.

Smith, Vernon L. 1994. "Economics in the Laboratory." *Journal of Economic Perspectives* 8 (1): 113–31. https://doi.org/10.1257/JEP.8.1.113

Sobel, Adam H. 2021. "Usable Climate Science Is Adaptation Science." *Climatic Change* 166 (1–2): 1–11. https://doi.org/10.1007/s10584-021-03108-x

Somin, Ilya. 2016. *Democracy and Political Ignorance: Why Smaller Government Is Smarter*. Stanford University Press.

Spence, Alexa, Wouter Poortinga, and Nick Pidgeon. 2012. "The Psychological Distance of Climate Change." *Risk Analysis* 32 (6): 957–72. https://doi.org/10.1111/j.1539-6924.2011.01695.x

Stevens-Rumann, Camille S., Kerry B. Kemp, Philip E. Higuera, Brian J. Harvey, Monica T. Rother, Daniel C. Donato, Penelope Morgan, and Thomas T. Veblen. 2018. "Evidence for Declining Forest Resilience to Wildfires under Climate Change." *Ecology Letters*. https://doi.org/10.1111/ele.12889

Stieb, Matt. 2019. "Poll: 12 Percent of Americans Have 'Never Heard' of Mike Pence." *New York Magazine*, 2019.

Stocker, Thomas F., and Andreas Schmittner. 1997. "Influence of CO2 Emission Rates on the Stability of the Thermohaline Circulation." *Nature* 388 (6645): 862–65. https://doi.org/10.1038/42224

Stoller-Conrad, Jessica. 2017. "Core Questions: An Introduction to Ice Cores." NASA. 2017. https://climate.nasa.gov/news/2616/core-questions-an-introduction-to-ice-cores/

Sunderrajan, Aashna, and Dolores Albarracín. 2021. "Are Actions Better than Inactions? Positivity, Outcome, and Intentionality Biases in Judgments of Action and Inaction." *Journal of Experimental Social Psychology* 94 (May): 104105. https://doi.org/10.1016/j.jesp.2021.104105

Tajfel, Henry, and John Turner. 1979. "An Integrative Theory of Intergroup Conflict." In *The Social Psychology of Intergroup Relations*, edited by William G. Austin and Stephen Worchel, 33–47. Brooks/Cole Publishing Co.

Tavoni, Alessandro, Astrid Dannenberg, Giorgos Kallis, and Andreas Löschel. 2011. "Inequality, Communication, and the Avoidance of Disastrous Climate Change in a Public Goods Game." *PNAS* 108 (29): 11825–29. https://doi.org/10.1073/pnas.1102493108

Thaler, Richard H., and Cass R. Sunstein. 2009. *Nudge: Improving Decisions about Health, Wealth, and Happiness*. Penguin.

Tingley, Dustin, and Michael Tomz. 2014. "Conditional Cooperation and Climate Change." *Comparative Political Studies* 47 (3): 344–68. https://doi.org/10.1177/0010414013509571

Tippett, Michael K., Chiara Lepore, and Joel E. Cohen. 2016. "More Tornadoes in the Most Extreme U.S. Tornado Outbreaks." *Science* 354 (6318): 1419–23. https://doi.org/10.1126/science.aah7393

Trump, Donald J. 2018. "Brutal and Extended Cold Blast Could Shatter ALL RECORDS—Whatever Happened to Global Warming?" Twitter. 2018.

Tversky, Amos, and Daniel Kahneman. 1974. "Judgment under Uncertainty: Heuristics and Biases." *Science* 185 (4157): 1124–31. https://doi.org/10.1126/science.185.4157.1124

Tversky, Amos, and Daniel Kahneman. 1986. "Rational Choice and the Framing of Decisions." *Journal of Business*. University of Chicago Press.

UNFCC, INDC to. 2015. "Press Statement: India's Intended Nationally Determined Contribution."

Urpelainen, Johannes, and Alice Tianbo Zhang. 2022. "Electoral Backlash or Positive Reinforcement? Wind Power and Congressional Elections in the United States." *Journal of Politics* 84, no. 3.

Vicens, Julian, Nereida Bueno-Guerra, Mario Gutiérrez-Roig, Carlos Gracia-Lázaro, Jesús Gómez-Gardeñes, Josep Perelló, Angel Sánchez, Yamir Moreno, and Jordi Duch. 2018. "Resource Heterogeneity Leads to Unjust Effort Distribution in Climate Change Mitigation." *PLoS ONE* 13 (10): e0204369. https://doi.org/10.1371/journal.pone.0204369

Vincent, Emmanuel M. 2017. "Scientists Explain What *New York Magazine* Article on 'The Uninhabitable Earth' Gets Wrong." *Climate Feedback*, July 12, 2017.

Visioni, Daniele, Giovanni Pitari, Glauco Di Genova, Simone Tilmes, and Irene Cionni. 2018. "Upper Tropospheric Ice Sensitivity to Sulfate Geoengineering." *Atmospheric Chemistry and Physics* 18 (20): 14867–87. https://doi.org/10.5194/ACP-18-14867-2018

Wallace-Wells, David. 2017. "The Uninhabitable Earth." *New York Magazine*. July 10.

Wallace-Wells, David. 2019. "Here's Some Good News on Climate Change: Worst-Case Scenario Looks Unrealistic." Intelligencer. December 20, 2019. https://nymag.com/intelligencer/2019/12/climate-change-worst-case-scenario-now-looks-unrealistic.html

Wang, X. T. 1996. "Domain-Specific Rationality in Human Choices: Violations of Utility Axioms and Social Contexts." *Cognition* 60 (1): 31–63. https://doi.org/10.1016/0010-0277(95)00700-8

Weber, Elke U. 2013. "Seeing Is Believing." *Nature Climate Change* 3 (4): 312–13. https://doi.org/10.1038/nclimate1859

Webster, P. J., G. J. Holland, J. A. Curry, and H. R. Chang. 2005. "Atmospheric Science: Changes in Tropical Cyclone Number, Duration, and Intensity in a Warming Environment." *Science* 309 (5742): 1844–46. https://doi.org/10.1126/science.1116448

Yamagishi, Toshio. 1986. "The Provision of a Sanctioning System as a Public Good." *Journal of Personality and Social Psychology* 51 (1): 110–16. https://doi.org/10.1037/0022-3514.51.1.110

Zaller, John. 1992. *The Nature and Origins of Mass Opinion*. Cambridge University Press.

Zefferman, Matthew R. 2014. "Direct Reciprocity under Uncertainty Does Not

Explain One-Shot Cooperation, but Demonstrates the Benefits of a Norm Psychology." *Evolution and Human Behavior* 35 (5): 358–67. https://doi.org/10.1016/J.EVOLHUMBEHAV.2014.04.003

Zimmer, Carl. 2013. "Bringing Them Back to Life." *National Geographic*, April 2013.

Zimmermann, Jarid, and Charles Efferson. 2017. "One-Shot Reciprocity under Error Management Is Unbiased and Fragile." *Evolution and Human Behavior* 38 (1): 39–47. https://doi.org/10.1016/J.EVOLHUMBEHAV.2016.06.005

Index